CENTIPEDES IN CAPTIVITY
The Reproductive Biology and Husbandry of Chilopoda

Orin McMonigle

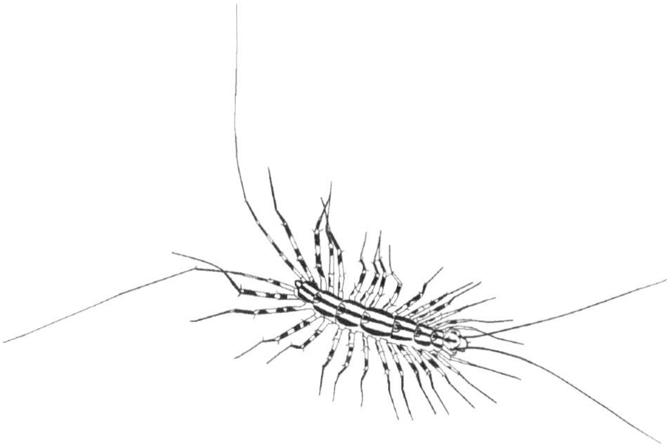

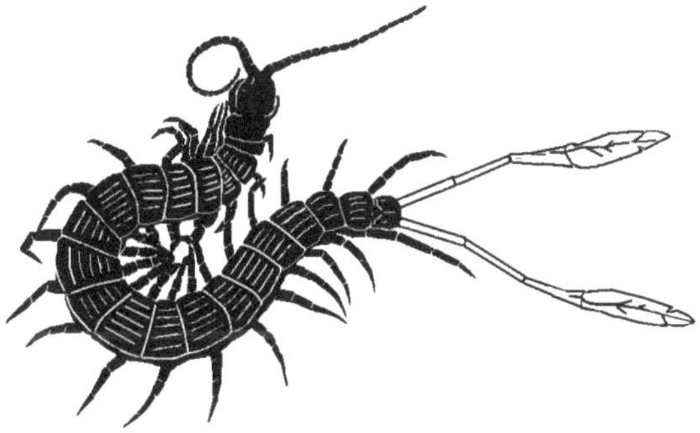

COACHWHIP PUBLICATIONS
GREENVILLE, OHIO

Scolopendra subspinipes, Bermuda. (CC) Sam Fraser-Smith

Centipedes in Captivity

ISBN 1-61646-231-0
ISBN-13 978-1-61646-231-4
CoachwhipBooks.com

Front cover: *Alipes grandidieri* protonymphs.
Back cover: (Top) *Scolopendra alternans* Haitian "Giant," large immature molt at thirty months of age, captive hatched. (Bottom) *Scolopendra dehaani* Orange Leg held with forceps.

Note: U.S. Geophilomorpha identifications are by Mercurio and are tentative based on the location and pictures only.

CONTENTS

ACKNOWLEDGEMENTS

My great thanks go to David T. Anderson for collaboration on associated articles and sharing personal experiences with centipede breeding and development. Special thanks is extended to Christian Schwarz, Randy Mercurio, Annie Lastar, Wade Harrell, Gillian Higgins, and Tammy Wolfe for editorial assistance. Great thanks go to John Skierski for husbandry experiences and text suggestions, as well as to Martin Gamache, Zhangjunduo, and Chia How for specimen details, also to Pierrino Mascarino for providing literature. Thanks to Dr. Rowland Shelley for taxonomic insight and information used in this text and the original handbook. Further input from Jim Smolka, Tom Larsen, Søren Rafn, Steven Lenaerts, and Nicholas Ditzel was much appreciated. Special thanks to Barney Tomberlin, David Moellendorf, Sam McAdams, Graham Criglow, and other collectors for *Scolopendra heros* subspecies range information. I thank my wife Sylvia for feedback on the design and for not noticing my live centipede collection. I thank God for providing the resources to put this book together and for creating these strange, wonderful animals.

DEDICATION

In memory of Robert Gale Breene III, Ph.D., who provided technical assistance on the 2003 centipede handbook and whose influence lives on in many areas of this text.

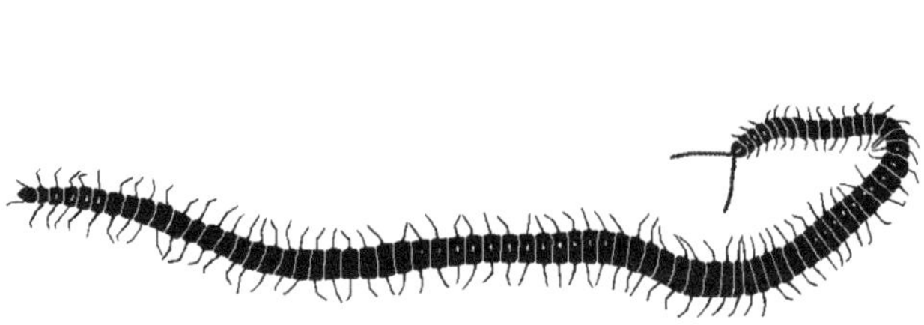

FOREWORD

THE MAJORITY OF centipedes sought in the hobby are large scolopendromorphs of the genus *Scolopendra*. Generally, these are powerful, venomous arthropods with lightning speed. Many centipedes could even be considered illusionists like Harry Houdini because they are capable of escaping what appeared to be secured enclosures. These qualities along with their aggressive hunting behavior and voracious appetites are perhaps what makes them so captivating to the hobbyist. Although some centipedes can be hostile and defensive there are many that are surprisingly very timid and reclusive. Regardless of their attitude, without the proper environment and captive care they can all quickly succumb to dehydration making them somewhat delicate creatures to keep as pets.

More than 10 years ago Orin McMonigle published *Giant Centipedes: The Enthusiast's Handbook*, which was essentially the only guiding resource available to hobbyists and dedicated to information on maintaining centipedes in captivity. Over the years, Orin has continued to share his experience and knowledge in papers such as the captive breeding of the African feather-tailed centipede (*Alipes grandidieri* (Lucas, 1864)) and the cosmopolitan house centipede (*Scutigera coleoptrata* (Linnaeus, 1758)). As the primary author, along with David T. Anderson, Orin published an interesting two-part paper on the captive hybridization of the North American centipede subspecies belonging to the nominal taxon *Scolopendra heros* Girard, 1853. More recently, Orin coauthored with Anderson regarding the development and husbandry of the orange- and black-banded Indian tiger centipede (*Scolopendra hardwickei* Newport, 1844). Contributions like these not only move the hobby in a new and improved direction but they advance our scientific knowledge of the little known and fascinating world of centipedes.

Now Orin's *Centipedes in Captivity: The Reproductive Biology and Husbandry of Chilopoda* extensively expands upon and compiles his previous works. Orin McMonigle presents us with a plethora of incredible photographs while summarizing his centipede adventures and know-how. We find that he also touches upon the smaller but still amazing centipedes of the orders Scutigeromorpha, Lithobiomorpha, and Geophilomorpha, which can be encountered by just about anyone that seeks them. I am confident that this book will help guide the new or even seasoned centipede enthusiast in caring for their centipedes properly and that it will also assist in preventing the common pitfalls of centipede husbandry. In addition, I expect this updated version to serve as the inspiration and guiding light for future centipede enthusiasts.

Randy J. Mercurio, 2014
Chilopodologist

Scolopendra sp., Australia. © Czardases

INTRODUCTION

CENTIPEDES INCLUDE some of the most massive and impressive arthropods on earth. Many species display incredible colors and striking combinations of red, blue, green, yellow, purple, orange, and black. Though they are magnificent living creatures, the plastic-looking body armor is so deeply pigmented with brilliant color some could be mistaken for high quality toys. Others have brightly colored alternative juvenile patterns or delicate developmental hues. The most colorful are usually among the largest, making it easy to see and appreciate the gorgeous patterns.

Beauty is not the only reason people are attracted to centipedes. Some of us feel a massive centipede's appearance of power is its beauty no matter the color. Others choose to keep centipedes in order to exhibit courage or confront irrational fears. I remember long ago when I would see a house centipede and cringe because I was told they had a painful bite, and they seemed to move like lightening. Yet now, with more experience under my belt, they no longer seem to move so fast. They are delicately handsome and immatures are pretty with purple hues, but aversion destroys appreciation. After keeping and enjoying them for so many years, when I see one running along my heart quickens a little with excitement rather than alarm. Not all fears are irrational since the larger centipedes can inflict a painful bite. Yet for some, fear brings attraction. The most massive *Scolopendra* are still intimidating to the experienced but they are not monsters to be dreaded, only timid and defensive animals to be respected. The notion of picking them up seems even more foolish after experiencing a painful envenomation, yet the biggest and most venomous species has never killed a human.

These spectacular animals have made their way into our homes as pets with regularity at least as far back as the late 1980s (Frye 1992). Large *Scolopendra* (often *S. subspinipes* or *S. dehaani*) have been a common sight for decades in specialty and large pet shops that carry a good selection of exotic animals. Yet for all this time, husbandry information for the keeper has been all but nonexistent. In recent years the Internet has become a useful tool but the beginner may not know where to look or which self-described expert's advice stems from successfully keeping centipedes for two weeks, two years, or two decades. This text is meant to provide practical information useful to the keeper in a hard copy form. The foremost goal is to present background, husbandry, and information on the common pet species. Small species are mentioned but the primary focus is on centipedes commonly maintained in captivity. Convincing captive specimens to breed is significant but secondary to keeping them alive and contained.

Scolopendra polymorpha "Aztecorum" immature from southern California. © John Skierski.

Malaysian "Beauty," 2003. Notice the right leg on segment fifteen is pale in color, a common result of regeneration. © Tom Larsen

Ethmostigmus sp., Tanzanian "Giant Yellowleg."

Otostigmus scaber female collected from Hawaii, USA. © John Skierski

Imports and available species have changed over the last few decades. The mid 1990s, when I began keeping centipedes, now seems to have been the heyday of centipede imports. At the time *Scolopendra gigantea* and the "Malaysian beauty," two of the most spectacular species on the planet, were pretty easy to get and not terribly more expensive than other available centipedes. Some amazing species like *Scolopendra viridicornis* were hard to find but not very expensive. Various African species and natives were also available and sometimes just for a few dollars. It could be argued that today is the new golden age of centipedes since many of the old standard species are available again, but at vastly inflated prices. Certainly, the beautiful *Scolopendra hardwickei* has joined the fold, but others have disappeared. Husbandry and reproductive biology are certainly better understood today. There will always be rare imports that contain unanticipated amazing creatures and it is hoped better understanding and documentation of their captive biology will lead to successful maintenance of stocks in captivity. Other than some beautiful *Scolopendra*, two of the more spectacular past imports were humungous black geophilomorphs that made our natives look like a piece of dental floss, and massive Malaysian Scutigeromorpha that eclipse our tiny adventive house centipedes.

Centipede books for the adult are very limited (though there are at least half a dozen children's books published since 2005). *The Biology of Centipedes* (Lewis 1981) and its reprints may be useful for the serious enthusiast, but very little relates to captive reproduction or husbandry since it is a technical text covering morphology, endocrinology, epidermal glands, natural history, etc. The earliest texts on keeping large-bodied centipedes in captivity may be a short section in *Keeping Spiders, Insects and Other Land Invertebrates in Captivity* (Murphy 1980) and the less than three pages in *Captive Invertebrates* (Frye 1992) with reference to *Scolopendra* (and a picture of *Scolopendra polymorpha* labeled "Copris texanum"). As appreciation for these spectacular captives picked up, some references began to populate general texts but the references had a few nice photos and little content. *Encyclopedia of Terrarium* (Bruins 1999) and *The Guide to Owning Millipedes and Centipedes* (Walls 1999) both offered a handful of color photos and a few pages of limited content. Quite some time later, *Invertebrates for Exhibition* (McMonigle 2011) offered a detailed chapter containing a few select species, some of those detailed here. The first dedicated book for the enthusiast, *The Giant Centipedes of the Genus Scolopendra: Their Captive Care and*

Husbandry (Sandefer 1998a) was a great beginning for hobbyists but it contained very large font text in a small format and lacked basic details. Its biggest flaw (beyond a few understandable taxonomic quaffs) was always the lack of availability even when it was in print. It has been out of print for a decade and it is nearly impossible to find a copy. The only text reliably available over the last decade from select invertebrate specialty vendors, *Giant Centipedes: The Enthusiast's Handbook* (McMonigle 2003), is limited to a small format and is starting to show its age. Since the original printing there have been advancements in husbandry knowledge, multi-generational breeding, and captive hybrids. Moreover, many novel ideas at the time related to sperm retention, mating, mites, and feeding have become common knowledge for today's enthusiast.

Other than *Invertebrates-Magazine*, centipede covers are exceptionally rare at best. 2006 cover.

Periodical literature is a hair more extensive than books, but still rather sparse. The *Bulletin of the Amateur Entomologists' Society* ran a short article on catching and keeping small British natives (Cloudsley-Thompson 1952), later reprinted and updated (Holford 2007). *Invertebrata* had a short spot on available pet centipedes from the genus *Scolopendra* (Mascarino 1996) and collecting *Hemiscolopendra* (Swift 1997). Breene's 1997 article in the American Tarantula Society's *Forum* tried to dispel some of the hype. Sandefer's *Scolopendra heros* articles in *Webbings* (1998b and c), no doubt timed to coincide with the book, may have included the first husbandry literature to concentrate on a single species. Ditzel (2002) covered reproduction and what was thought to be the available knowledge at the time of *Scolopendra subspinipes* subspecies in *Exotiske Insecter*. The only relatively extensive periodical literature on centipede husbandry has run in *Invertebrates-Magazine* by myself (2006, 2008, 2014), Anderson (2008) or both of us (2011a, 2011b, 2013). Most of the detailed breeding and husbandry information is related in this text. Scientific journals have run countless articles running from charting how fast centipedes die from drying out, to injecting small living vertebrates with concentrated venom, and of course naming new species and changing names. Those items of interest to the centipede keeper will be discussed.

Centipedes are seen in popular media but are surprisingly rare when compared to the familiarity of the animals themselves. Unlike many of their invertebrate brethren, everybody knows what a centipede is (notwithstanding a little milliped and worm confusion). There are a few seconds to a few minutes on National Geographic shows, the BBC's *Wild Things* (2013) and various bug war shorts. The most famous documentary footage is probably the

Scolopendra dehaani, standard coloration for Vietnamese imports.

two minutes that include a South American *Scolopendra gigantea* climbing a cave wall, hanging by the rear sets of legs, and catching a bat from midair in Attenborough's *Life in the Undergrowth* (2005). A tribute more recently followed in the "Giant Centipede" episode of *Wild Things* (2013), a one-hour show dedicated to this group of fantastic arthropods, specifically hunting down the bat-eating *Scolopendra gigantea* in Venezuela. In popular television, *Marvel's Agents of S.H.I.E.L.D.* (2013) premiered with agents fighting Project Centipede, but it was only a piece of alien technology that looked a tad centipede-shaped. On the big screen there are various science fiction movies with centipede-like creatures but nothing that stands out. The horror film list is short and notable only for its lack of quality. There's *Centipede!* (2004) about some Americans exploring an Indian cave full of deadly centipedes and an Asian language low budget horror movie with a ton of live centipede footage called *Centipede Horror* (1984). The only film you might find at a local video store is a pair of nasty and crass B movies (*Human Centipede* and its sequel) which glorify human torture and have nothing to do with Chilopoda (plus each "segment" has four appendages like millipeds, not centipedes). There are some short cameos in movies such as *Indiana Jones and the Temple of Doom* (1984). A centipede shows up in one scene in the first *Lord of the Rings* movie and a couple of giant CGI centipedes take the stage for a few seconds in Peter Jackson's *King Kong* remake (2005). The 1991 movie *Naked Lunch* has quite a few centipede cameos, both real and special effects. In the movie, "centipede meat" is consumed as a narcotic. Early on the main character Bill Lee, an exterminator, encounters a live *Scolopendra* in the bathroom, and later dried centipedes on sticks are seen being bought and sold as drugs. Another scene shows a butcher cutting up a giant (presumably latex) centipede and one character transforms into a half man, half-centipede creature and does something very nasty to one of the other characters. It is a disturbing movie (basically meant to be a long drug hallucination) but not nearly so base or appalling as the concept behind *Human Centipede*. When it comes to video games it is hard to imagine anyone is unfamiliar with the classic arcade *Centipede* (1981) or one of its many relaunches (its original sequel however was *Millipede* (1982)). *Escape From Bug Island* (2006) requires the smashing of hundreds or thousands of cat-sized house centipedes among other giant "bugs." Last and least of all, there might be more plastic centipedes in landfills than those presently alive on earth.

SPECIMEN IDENTIFICATION AND TERMINOLOGY

Anyone who keeps centipedes is likely to learn at least a little about their taxonomy and identification. Many are referred to only by their scientific names since one can only say *tiger* in so many ways. Some species are not so perfectly understood due to variation between specimens or populations, others are undescribed, while the binomial for some can seem more fluid than common names. Even so, incorrect species names can be the result of simple error more than taxonomic difficulty. Confirmation of identification is often possible by looking at a few easily seen features. Some subspecies have different morphological characteristics (beyond color which means nothing when inconsistent) from the nominate species, which suggests interbreeding is unlikely in natural or laboratory settings. Other specimens vary in color and size but could be entirely unrelated or the same species. Here we will try to get the beginner up to speed. Some basic indications as well as specific features for common species are arranged in later chapters.

The original 7-tier system and binomial nomenclature has changed. Suborders, superfamilies, superclasses, subclasses and tribes are common to some other arthropod taxonomic schemes, but usually are not used for centipedes. This text will only delve lightly into subfamily and subspecies beyond the standard tiers. Massive databases could be filled with taxonomic changes, rules, and information. This book is intended as a guide for the enthusiast and includes taxonomic information commonly encountered by those interested in studying, observing, recording, and enjoying the development, lifespan, and reproduction of living centipedes.

Despite difficulties with taxonomy, scientific names are helpful to the hobbyist and in some cases they may be more reliable and usable than common names when buying and trading centipedes. Of course there has to be more than a little thought in the original identification (every so often a *Scolopendra subspinipes* from Asia is seen labeled as a South American *Scolopendra gigantea*). Some vendors may be looking for the sale, such as an e-mail I get every so often selling *S. gigantea* from Africa, but most are good-hearted attempts at putting a name to their animal. Recently, at a small conference, I stumbled upon a specimen in a tub labeled *Ethmostigmus rubripes* Australia. I was more than thrilled because I had never seen a live Australian centipede before. I asked to take some photos and spoke to the vendor. From the spiracles I immediately realized it was a *Scolopendra* and recognized shortly afterwards it was just a pretty standard *Scolopendra polymorpha* from the southwestern U.S. I found out the vendor had googled pictures and picked out one with a similar color pattern. He seemed glad to get a better identification.

Scolopendra heros heros banded from west Texas.

Tanzanian "Neon Blue" adult pair.

My 1994 centipede acquired as *Scolopendra gigantea*. It was either identified correctly or a very large *Scolopendra galapagoensis*, but I no longer have the specimen, just a few poor 35mm prints.

Notice the various grooves in the body armor. The shallow grooves that commonly divide the upper body plates into roughly three sections are often termed paramedian grooves or sulci (plural for sulcus). Paramedian means near the middle. There are also two very shallow grooves in the headplate but the right one is obscured by the flash. Finally, the large deep groove across the first tergite is commonly known as the ring furrow. (It may be seen labeled as anterior transverse suture though it is crescentic, not transverse, and anterior refers to the placement on the tergite, not the location of the tergite.) The presence or absence of any of these grooves can make or break a species identification, but only the ring furrow is easy to see in most photographs. This is a very young, captive-bred *Scolopendra gigantea.* (The smooth antennomere count is lower in immature specimens.) © David T. Anderson

Every so often some specimens have come up labeled with cf., short for *confer* or *compare to*. An example would be a recent appearance of the "*Scolopendra* cf. *gigantea*" label on obviously unrelated specimens. However, cf. is not supposed to be used for uncertain or doubtful identification but rather for uncertainty in the taxonomy. This means a large *Scolopendra* from Hispaniola with no ring furrow, lacking 3-5 dorsal prefemoral spines on the legs, and with five smooth antennomeres should never be called *S.* cf. *gigantea*. On the other hand, a known population from South America believed to be *Scolopendra gigantea,* but with some very slight differences that cross over with those in the species description of its closest relative *Scolopendra galapagoensis* might employ cf. Usually it would be labeled one or the other, or *Scolopendra* sp. Some of the specimens pictured here are closer to *S. gigantea* than *S. galapagoensis* and will be labeled as the species description they are closest to rather than be confused with the so-called "cf. gigantea."

Closest to *Scolopendra gigantea.* © David T. Anderson

Fortunately, the characteristics which differentiate suborders and species have become more accessible to the inquisitive enthusiast, although the majority of papers require institutional membership or extensive monetary outlays for access. Even after the literature has been purchased, discovering these features is not always easy since older descriptions can be vague, lack basic details, or differ from other literature. Important morphological characteristics such as hirsute antennomeres and spines on the legs and spurs are somewhat to highly variable within a given species (or the same specimen by age) though the standard is generally not so variable. Many new centipedes in the hobby have not been identified and are commonly referred to as *Scolopendra* sp., followed by the name of the collection location. This is not

Scolopendra heros arizonensis feeding on a slice of "beetle jelly."

a reliable method of identification because the centipedes may have a wide distribution, a number of different species can live in the same area, and it may not be the correct genus. Scolopendromorpha species can look very similar even across genera or can look very different and still be the same species, especially *Scolopendra polymorpha*, but there are even forms of this species that seem to conform to a unique animal and yet fall well within the standard variation of characteristics. Although many wild species can be positively identified to the species level, there are quite a few that may never be.

When it comes to some of the species and varieties that had been kept and bred over the last few decades, sometimes a key feature in a photograph can confirm the old identification or lead to a new one. However, we may never know what species many of them were, even in cases where identification should be easy because the photographs rarely show important features used in identification. Moreover, specimens are rarely preserved or are not easily accessible if they have been preserved. Enthusiasts should attempt to preserve unusual or unidentified specimens through drying, freezing, or placing in 75% ethyl alcohol or isopropyl alcohol immediately upon the death of the animal. Of course a written record of the specimen's date of acquisition and import location are minimum requirements. (A bunch of unlabeled centipedes in freezer bags is a waste of space.) These can be saved until better access to defining details or new descriptions have been made for identification. Sending these to a taxonomist is one option, but unless the province or state of the collection location is known it may just end up in the trash. Physical features are not always enough.

Describer names and the year of description are included within the species lists to give context to the names since the person and year can carry interesting details in themselves. These are not meant to be part of the scientific name and these extra details cannot change the validity (or lack) of a species identification. As standard, those for which the genus has been changed after the original description are in parentheses.

The technical term centiped is not used in this book. Unlike the diplopoda (the millipeds) there is almost no use of the technical term even within the scientific literature.

Terminology here will follow commonly used words as much as possible. For example, instead of glabrous (meaning bald, hairless, or smooth) the word smooth will be employed. The terms groove or furrow will be used for sulcus and suture. (Sulcus and suture both refer to a shallow groove in the plates of the exoskeleton, sometimes used interchangeably or varying arbitrarily according to depth by author.) However, there are no common words for coxosternum, maxillary palp, etc. A glossary is provided at the back of this book to bring the beginning hobbyist up to speed.

THE INTENT OF *CENTIPEDES IN CAPTIVITY*

In the late 1990s it seemed more and more invertebrate books would contain at least a page or two on centipedes but there has not been much further progress. Unfortunately, the only in-print, English-language book devoted to the care of centipedes is *Giant Centipedes: The Enthusiast's Handbook* which is limited in scope by its construction and was written more than a decade ago. This text is an exponentially expanded and updated version of the handbook and includes adaptations of various articles on centipedes that have been published in *Invertebrates-Magazine* from 2006 to present. This book is likewise written to promote the centipede hobby and offer information not readily available to the hobbyist, presented in a large hardcover format and in full-color.

The chapters discussing the five centipede orders are listed in the order of their captive efforts. Scolopendromorpha is the longest and most detailed chapter and the primary interest of most enthusiasts. Despite the extreme paucity of obtainable species, the Scutigeromorpha have seen some captive efforts and successes. The Lithobiomorpha are commonly captured and kept by the new and young enthusiast but breeding efforts are scarce. Although there have been some fascinating Geophilomorphs in captivity, their full life cycle has rarely been followed—hatching out some babies which die before reaching maturity offers little information of value. Craterostigmomorphs have not been reared in captivity and early development has not been documented despite the way it is often reported.

The goal of this book is to present all the available information in a format that is worthy of these cryptic and eerily beautiful predators of the dark. The goal is to combine a husbandry manual, a large picture book, and a biological textbook tailored to the advanced enthusiast, but still readable by the beginner (so that a normal person enthralled with the fascinating centipedes might read a few pages or more without passing out cold or swearing never again to open a book on invertebrates). Hopefully the availability of useful husbandry details will encourage expanded captive-breeding efforts, research, and interest. Large centipedes can be excellent display specimens and easy captives though they are certainly not handleable "toy" pets. This text is geared to provide detailed information for the successful husbandry and understanding of the amazing centipedes.

RELATIONSHIPS, MORPHOLOGY, AND BIOLOGY

THE CENTIPEDES are a group of primarily predatory arthropods in the class Chilopoda. There are at least 3,150 described nominal species in five orders and twenty-four families. Approximate known species for three of the orders below are according to Zhang (2011). Britain hosts forty-seven species including three from the order Scolopendromorpha in the genus *Cryptops*, but not including seven found only within buildings or heated greenhouses (Lee et al. 2007). Including established introduced species there are only twenty-one scolopendromorphs in North America (Shelley 2002) though the number could be greater than thirty including those intercepted in quarantines. According to Mercurio (2010) there are ~7 scutigeromorphs, ~375 lithobiomorphs, and ~211 geophilomorphs known in the fifty United States, Canada, and Greenland, which includes introduced species that are established or of uncertain establishment. Due to limited work the fauna of many other regions is even less completely studied.

- Phylum Arthropoda
  - Class Chilopoda
    - Order Scolopendromorpha
    - Order Scutigeromorpha
    - Order Lithobiomorpha
    - Order Geophilomorpha
    - Order Craterostigmomorpha

A typical scolopendromorph, *Scolopendra polymorpha* from California, USA. In many of the California forms of this species, the body, head, and last tergite are uniform in background color. This specimen is an adult male which lived three years in captivity and molted twice without growing, ~3" (7.5 cm).

The order Scolopendromorpha includes approximately 678 species. All the common pet species are in this order. Smaller species are a little over an inch while the largest (*Scolopendra gigantea*) commonly attains a length of 10.5" (27 cm). Most of the familiar giants (> 3" or 7.5 cm) come from the genus *Scolopendra*, though members of the genera *Alipes*, *Ethmostigmus*, and *Otostigmus* from Africa and

Cormocephalus sp., Madagascar.
© Frank Foldbjerg Hansen

Cormocephalus from Madagascar and New Zealand, among others in the Family Scolopendridae can exceed 4″ (>10 cm). The single *Plutonium* species from southern Europe, an oversized representative of the Family Plutoniumidae, can grow to 6″ (15 cm). These large centipedes are primarily predatory and amount to the only order capable of regularly taking vertebrates as prey. Most scolopendromorphs have forty-two or forty-six legs and segments, while only representatives of the Family Scolopendridae have clusters of four ocelli on both sides of the head. Mimopidae have one pair of ocelli and the remaining families have no eyes (Lewis 2006). The rear legs are large, and can be adapted to function as pincers, rear antennae, colorful warning flags, or for making noises (stridulation). Females are known for their maternal care and look after the eggs and young until second instar. Broods range from 20 to 156. First instars begin life with a full complement of legs.

Scolopendromorpha eggs are protected from the environment by their mother and therefore have extremely thin and fragile shells. They are yellow because the shell is transparent. © David T. Anderson

Scutigeromorpha eggs have a thick, white-colored, leathery shell and can be handled easily without damage, *Scutigera coleoptrata.* After hatching, the shell does not disintegrate and can remain intact for years even on damp substrate. Note size compared to wingless *Drosophila melanogaster* fruit flies near bottom of photo.

The order Scutigeromorpha contains about ninety-five extant nominal species (Minelli 2011) in three families, all with long antennae and compound eyes. Adults have thirty extremely long walking legs but just eight visible tergites. The rear legs resemble antennae and are usually longer and thinner than the antennae. These species are the exception to many centipede "rules." Some of them are day active and none are capable of burrowing. All species lack the extensive trachea which bring oxygen to every cell in the other orders, but instead use blood to get oxygen to most cells. Every leg is long and the tarsi are hypersegmented, so

This golden beauty is a native house centipede found only in Arizona, USA, *Dendrothereua homa* (three legs have been autotomized). © John Skierski

much that each foot has a greater number of small segments than the antenna of most members of other orders. They feed primarily on small, active insects (rarely arachnids or other centipedes) and the largest are unlikely to attack the smallest of hatchling vertebrates. The body length of the common house centipede, *Scutigera coleoptrata*, can be a little over an inch (3 cm). Most hobbyists have this centipede in their homes, but not often as pets. The house centipede is often unnoticed because it lives in dark areas, is seldom found in numbers, and runs away with improbable speed. Specimens have been clocked at 1 foot, 5″ per second (>42 cm/s) (Manton 1952). An arboreal species encountered in India (probably the largest species in the Order) can grow to nearly 4″ (10 cm) in body length; it is an expansive animal when the long legs are considered and the land speed it clocks may be proportionately impressive. Eggs are laid individually in dirt and cracks. First instars begin life with eight legs, reaching the maximum of thirty legs only after six molts and maturity after a dozen or so.

The order Lithobiomorpha is a massive order with species that look very similar to each other. There are two families with approximately sixty-three genera containing 1,120 described species. Many resemble small or young Scolopendromorpha but they have only thirty legs, each ocular field can contain up to three dozen ocelli, and alternating body segments are reduced. There are fifteen corresponding tergites, though some overlap greatly.

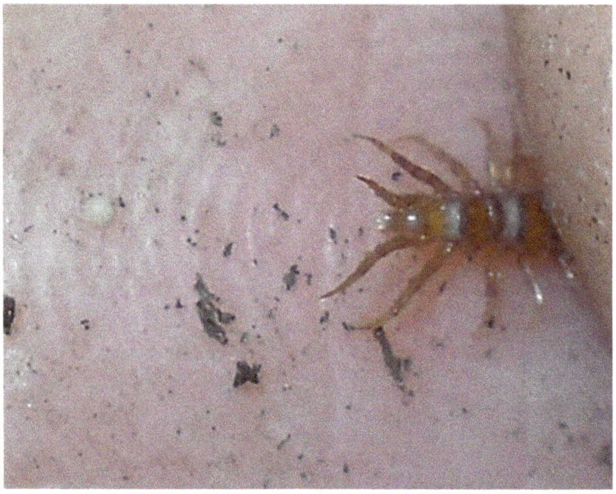

This dwarf lithobiid, *Tidabius tivius,* egg is smaller than the distance between the rings of a fingerprint. Egg shells in this order are so sturdy even the smallest eggs can be handled. This ovum was manually removed from the mother`s ovipositor and hatched five days later. The shell feels like it would be as sturdy and long-lasting as that of scutigeromorphs, but the first instar lithobiomorph eats it.

The rear legs are usually enlarged but rarely specialized to any great degree. The adults range from 1/4" to 1 1/2" in length (6-38 mm). They appear to be generalist omnivores in captivity but this may relate to specific prey uncertainty. Females deposit eggs in the soil individually, or in batches up to twenty. First instars start out with fourteen legs and develop thirty legs following the third molt (fourth instar). They continue to grow and require a number of additional molts to reach sexual maturity.

The order Geophilomorpha contains at least 1,254 known species ranging in size from 1/2" to 10" (1-25 cm). Most do not exceed 2 1/2" (<6 cm). The huge ones (though threadlike and much less massive than a *Scolopendra* of the same length) would make interesting captives but have not been available. The body is long, thin, and made up of at least thirty-five segments.

Lithobiomorpha. *Lithobius forficatus* from Medina County, Ohio, found under a limestone planting border.

Geophilomorpha often wind around soil particles so it is difficult to make out where one animal ends and another begins. *Arenophilus bipuncticeps* from northern Ohio, USA.

The legs are often shorter than the width of the body. This is the only order in which the terminal pair of legs often seem to be undifferentiated but there are species that run backwards with long terminal legs held up like antennae and many have inflated or oddly shaped terminals. Even the largest species feed on small food including tiny mites, springtails, and nematodes (the alternate name "worm centipedes" refers to their appearance rather than diet). Some feed on decaying vegetation while the European *Haplophilus subterraneus* sometimes eat live plants (Nichols et al. 1971). The geophilomorphs are blind and found under leaf litter or deep in the soil. Nearly all live in the ground; *geo* = earth and *phil* = love. However, some species live under rocks in tide pools or inhabit sandy beaches and mud flats. Some abandon eggs in the soil but the majority are known to brood eggs like Scolopendromorpha. First instars have the same or nearly the same number of legs as the adults.

There is only one family, one genus, and two described species in the order Craterostigmomorpha. One is found in Tasmania and the other in New Zealand, but otherwise they are quite similar. Members of this order have thirty legs and superficially resemble the Lithobiomorpha except they appear to have twenty-one tergites because some are divided into two sections. Mothers watch over their eggs and young, which hatch with a reduced number of legs.

The relationship between the different orders is not perfectly understood. The Craterostigmomorpha are usually considered closest to the Lithobiomorpha but others considered them more closely related to Scolopendromorpha (Lewis 1981) because they look after their young. The Scutigeromorpha had been combined with Craterostigmomorpha and Lithobiomorpha because they have unevenly formed tergites, more articulate maxillipeds, and anamorphic development. (Larvae start out with a smaller number of legs and segments which are added only as they progress through molts.) All representatives of these three orders have the same number of legs (thirty) at maturity. Geophilomorpha and Scolopendromorpha share epimorphic development (larvae usually start out with a full complement of legs and body segments) and there are no reduced tergites in between larger or overlapping tergites. The epimorphic centipedes share a similar development stage with most millipeds and arachnids. This stage precedes the first instar and is typified by underdeveloped appendages and lack of mobility. In millipeds and centipedes it is known as the protonymph and in arachnids the postembryo. More recent schemes discount developmental and structural similarities and place the Scutigeromorpha by itself (Subclass Notostigmophora) because of the unique dorsal spiracles and unusual respiratory system. The remaining four orders make up the Subclass Pleurostigmophora. Understanding the closeness of relationships can be compounded by the lack of ability to prove or understand which features are truly "primitive" or derived such as compound eyes, reduction in articulation, and greater or fewer body segments. Other barriers seem related to the tendency to throw out coding genes which do not support an hypothesis and misunderstood or lacking biological details such as the misconception that epimorphs regenerate "the last legs only or not at all." (Edgecombe and Giribret 2007).

GENERAL CONSTRUCTION

The centipede body is divided into two sections, the head and trunk. The head contains two conspicuous antennae, a chewing mouth, and sometimes eyes. The trunk has anywhere from six segments (four leg pairs, maxilliped segment, and growth segment) in first instar Scutigeromorpha (McMonigle 2008) to the 191 segments of some mature Geophilomorpha (Chipman et al. 2004). The first segment of the trunk contains the maxillipeds which wrap forward under the head and from below can appear to be part of the head. Venomous maxillipeds are unique to centipedes and the large, often toothed, plate formed by the base of these legs describes the class (*chilo* = lip, *poda* = foot). The remaining segments also each hold a single pair of legs. The upper surface of the body is covered with hardened plates called tergites and the relative size, fusion, or segmentation can be different among orders. The underside of each segment is covered by a hardened plate known as a sternite. The last segment contains the terminal legs, genitalia, and anus.

ANTENNAE

This pair of sensors is the most important single sensory apparatus of the centipede. Each antenna is made up of more than a dozen to hundreds of small segments known as antennomeres. The antennae are coated in sensory

The "fuzzy" antennomeres of the large scolopendromorphs are coated in microscopic setae as seen here. Each one is connected to a nerve cell. Scanning Electron Micrograph, *Scolopendra dehaani*. © Travis Knudtson

The antennae are held in all sorts of strange configurations commonly only noticed in photographs. *Scolopendra polymorpha*.

Here it is very easy to see the first six antennomeres (from the base) are smooth, while the last eleven appear "fuzzy" or hirsute because they are coated in a dense mat of tiny setae (hairs, bristles). The right antenna is being cleaned by the maxillary palps. *Scolopendra dehaani* ex. Vietnam.

Blue Ring Centipede with curled antennae. © John Skierski

Lithobius forficatus have larger setae on the antennae, but they are much smaller animals. Each antenna of this specimen is composed of 42 antennomeres.

setae ("hairs") and are incredibly articulate. In Scutigeromorpha the antennae are incredibly long, thin, and composed of large numbers of segments. They usually resemble assassin bug antennae due to an articulated joint. (They can be moved from the base and towards the middle.) Scutigeromorph antennae are not the only ones to exceed a hundred antennomeres. Some of the troglobitic lithobiomorphs such as *Lithobius matulicii* from caves in Bosnia and Herzegovina have antennae with more than 100 segments (White and Culver 2012). Geophilomorphs always have fourteen antennomeres while Scolopendromorphs usually have from seventeen to twenty-seven.

In many species the antennae are massive and thickened, and used like fingers to probe objects. The setae are very sensitive to touch and, with few exceptions, only dead animals do not respond if the antennae are touched or blown on. Reaction to smell is difficult to observe or test but direct contact with an antenna is enough for an object or prey to be recognized as food. Unlike those of insects, centipede antennae never seem to be adapted for primarily olfactory purposes because none have long, feathery outgrowths or massive setae and they are not sexually dimorphic (one sex does not have larger antennae to locate the other by pheromone).

Setae are the basic sensory apparatus of the antennae. In very small species the setae can be relatively large, but functional setae can only be so small. Tiny setae can appear large on very small arthropods. In the large species there are usually impressive sheets of nearly microscopic setae on the antennomeres that look like a fuzzy coating, especially near the ends. Basal antennomeres can have very few setae and so may be called smooth or glabrous (which means smooth, bald, or hairless). Since

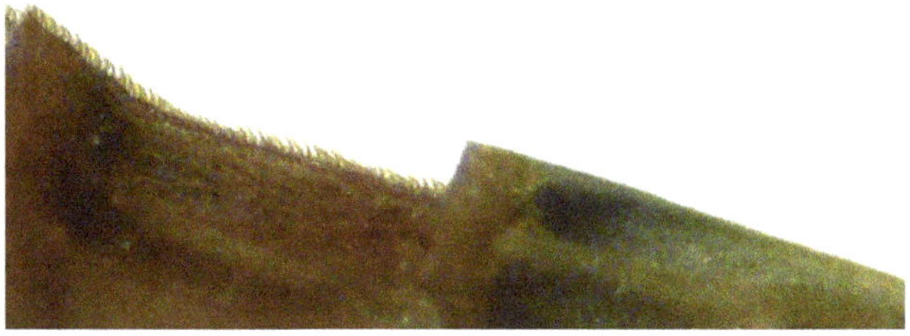

Smooth and hirsute antennomere surfaces. *Scolopendra alternans* dried specimen.

even the smooth segments may have some microscopic setae, the term sparsely hirsute (hirsute meaning hairy) is sometimes used, but such terminology is akin to arguing there can be no bald men, only sparsely hirsute ones. The number of smooth segments has long been used in classification and identification schemes though it can vary for subspecies, populations, and individuals, and even can vary for the same individual at different ages.

Forward from the base of each antenna is found a Tömösvary organ in centipedes with exception of the epimorphic Scolopendromorpha and Geophilomorpha. These nearly

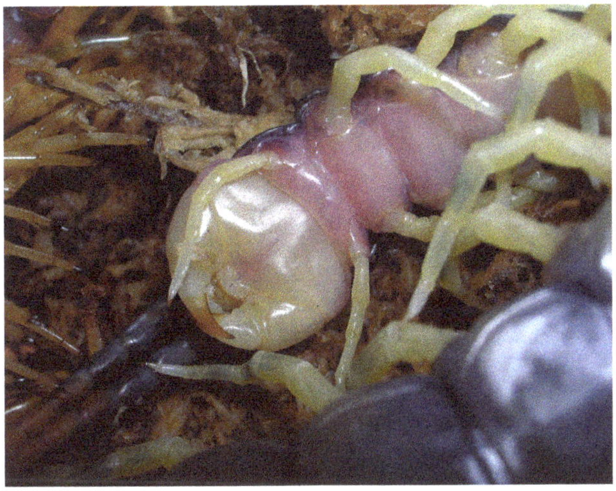

Maxillipeds and coxosternum, *Scolopendra gigantea*. © David T. Anderson

The leg-like second maxillary palps and the flattened first maxillary palps are easy to see here behind the maxillipeds. *Scolopendra dehaani.*

Scutigeromorph second maxillary palps are spiny, long, and easily seen from above.

microscopic pits are sensory organs of uncertain function that are also found in millipeds, symphylans, and springtails.

MAXILLIPEDS

The prefix "maxilli-" refers to the mouthparts of centipedes and insects (maxillae) which are located behind or around the mandibles and usually end in a spine. The suffix "-ped" means foot. These prey-catching forelegs are what truly make centipedes unique. Chilopods are the only invertebrates with legs that contain venom glands and whose claws end in needle-like points to deliver the venom. Centipedes do not truly "bite." Although they give a venomous pinch or hug, nobody tells stories about the most excruciating centipede hug they have experienced.

These highly adapted legs are considered a fourth set of mouthparts, despite origin, and have traditionally not been added to leg counts. (All the anamorphic centipedes would have 16 pairs of legs rather than 15 were maxillipeds included.) Although it is unlikely, knowledge of the true origination of the maxillipeds could be what has led to myths about walking legs being capable of delivering venom. All antennae, mandibles, and maxillae in arthropods are believed to be derived from legs, so maxillipeds as such are actually the origin of these organs. However, in chilopods, the maxillipeds have secondarily developed from an already derived walking leg.

Maxillipeds themselves are not unique since crustaceans including every crayfish, lobster, crab, and isopod also have varyingly adapted front legs used as an extra set of mouthparts. Venomous maxillipeds are unique.

Despite uniqueness, they have been awarded a whole slew of different names including forcipules, gnathopods, and prehensors (as well as poison claws and chelicerae among others). In this text I have chosen to stick with standard arthropod terminology rather than try to argue the various trains of thought involved in picking the "correct" alternative specific suggested requirement.

In most species it is very easy to see the maxillipeds arise from the first segment behind the head though they are much thicker and shorter than the other legs. The massive maxilliped claw is also known as the tarsungulum. The tarsungulum is greatly enlarged, heavily sclerotized, and there is a small opening near the tip connected to the venom gland. The gland extending into the basal segments is surrounded by muscles which propel the fluid during a "bite." The maxillipeds must be greatly thickened to contain venom, glands, and muscles for injection and movement.

The maxillipeds project from either side and are curved inward and forward. They are not only thick, but often shorter, and not articulate like other legs. Still, they can be surprisingly gentle, especially in the case of the extreme tenderness required to manipulate the thin-shelled eggs during cleaning. Some centipedes, especially *Scolopendra*, also use their maxillipeds to test and probe objects, but the movement is provided by movement of the front of the body since the maxillipeds only open and close. They come together like a pincer, or like the jaws of a beetle or praying mantis. In the epimorphs like *Scolopendra*, the base has the same number of short, squat segments but each maxilliped moves as a single piece with only two points of articulation. The basal segments of the anamorph legs are more articulate, especially in the scutigeromorphs, but they still do little more than open and close. There is very little range of movement. The primary purposes of these appendages are to envenomate and hold tightly onto prey—maxillipeds do not chew. The prey is often held close to the mouth for chewing, or carried with these front legs to be set down and chewed up in private. They are similar in function to the raptorial forelegs of praying mantids or the pedipalps of whipspiders (amblypygids). However, maxillipeds are also used to move or carry away large objects when digging, like the multifunctional pedipalps of the scorpions and whipscorpions (vinegaroons).

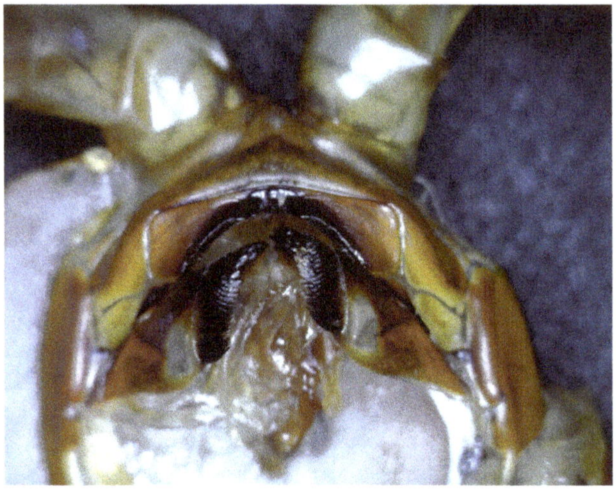

The chewing mandibles are only visible after the first maxillae are removed. The hardened edge above is the upper lip or labrum. *Scolopendra gigantea* older immature. © David T. Anderson

The lower or basal surface of the maxillipeds is a large plate known as the coxosternum. As mentioned, this is the "lip foot" of the Chilopoda. It is thought to be the sternite of the first body segment fused to the coxae of the maxillipeds, but it covers much of the bottom surface of the head (each side is termed a coxosternite). The coxae are not fused in scutigeromorphs and there are four or more long, thick setae facing the mouth. In other orders there is usually a small row of immobile teeth at the leading edge of the coxosternum which helps to hold food items in place against the opposing mouthparts. The number and shape of these teeth are useful in some identifications.

MANDIBLES AND MAXILLAE

There are three sets of mouthparts used to orient and chew the food. The first maxillae are short and thick and used to grind the food and pull it towards the sharp chewing mandibles. The first maxillae look like a pair of small, white molars placed sideways because the segments lay flat against the bottom of the mandibles. The sharp, chewing mandibles are located beneath the first maxillae and usually cannot be seen on a living animal except during feeding. The front edge of each mandible has sharp cutting teeth while the bottom or rear edge is outfitted with a thick brush of hairs (setae). The chewing edge (teeth and setae) are highly sclerotized like the claws and are normally dark brown to black in color.

Outside the mandibles and the first maxillae are the second maxillae which terminate in the maxillary palps. This third set of mouthparts has the same segmentation as the legs (coxa, trochanter, prefemur, femur, tibia, and tarsus). The coxa and trochanter are widened plates while the remaining segments (the palps) are long, finger-shaped, and look much like a tiny leg. The tarsi of the palps have a large inner, brush-like surface used for cleaning the antennae during grooming. The palpi are manipulative appendages also used to orient and direct the food towards the first maxillae and mandibles for chewing. The palps can have more or less spines and teeth depending on the genus or species.

Unless the food is too large or too difficult to chew, every piece is pushed down the centipede's throat without a scrap remaining. The mouthparts are impressive tools for shoveling in material but the orientation makes them useless for chewing through hardened objects, even rather thin items. One collector sent me large *Scolopendra* from Arizona inside zip-lock bags for years and not one ever made a hole (if any had, the bags were inside double-walled containers).

Many species have no eyes, such as this common Scolopendromorpha, *Scolopocryptops spinicaudus* Wood, 1862 from western Washington. © John Skierski

EYES AND PREY RECOGNITION

There is some variation among related species but eye structure is primarily determined by the order or family. Members of the Scolopendridae in the Scolopendromorpha normally possess four tiny ocelli on either side of the head while the Mimopidae have a single pair and the other families have no eyes. Geophilomorpha never have eyes while the Craterostigmomorpha have one ocellus on each side like some members of the other orders. Lithobiomorpha usually have eyes, those in the Family Henicopidae (about 10% of known species, mostly tropical) again have a single pair of large ocelli while those in the much larger

Scutigeromorpha eyes can have incredible numbers of facets. Philippine species. © Martin Gamache

The Scolopendridae have four eyes on each side of the head capsule. *Scolopendra dehaani*.

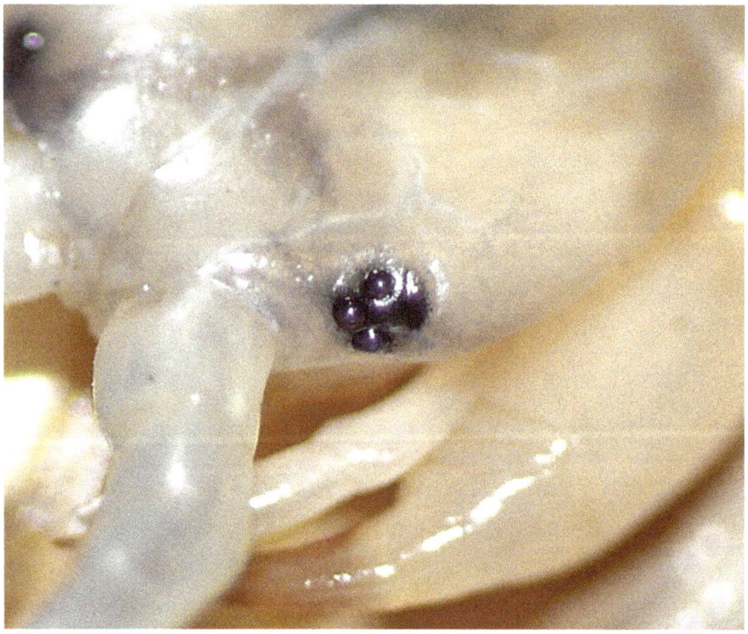

Eyes of *Scolopendra viridis*.
© John Skierski

Family Lithobiidae seem to have a pair of tiny eyes. However, each of these "eyes" is actually a cluster of less than a dozen to more than three dozen ocelli within a small ocular field. The ocelli are tiny but easy to see with a 10x loupe. The Scutigeromorpha always have advanced compound eyes and in some species the total number of facets or ommatidia exceeds 1,000 (individually in excess of 500 ommatidia).

Visual prey recognition is rare in centipedes. More than half of all known species are blind (although this is mostly the Geophilomorpha). Even *Scolopendra* with easily seen lenses virtually ignore prey until it comes into contact with a part of the centipede. Only then do they attack and coil around the prey. They may be able to detect prey vibrations with their antennae, but if so, the prey would need to be extremely close. Small stone centipedes with multiple ocelli also seem unaware of prey at any distance. The house centipedes, just as the large compound eyes would suggest, can target, chase down, and catch prey a few inches or more distant (further away than the extremely long and thin antennae). Although primarily ambush hunters, they are capable of targeting prey, specifically flittering moths drawn to lights at night, a foot or more distant.

Centipedes are nocturnal and use their antennae and legs (including the maxillipeds) to evaluate the environment and get along quite well with very tiny or no eyes. They do not have any of the unique systems to detect vibrations like many arachnids (such as pectines, malleoli, or trichobothria), but there are a number of setae that make them responsive to air movements and sounds loud enough to cause vibrations. *Scolopendra* often react if yelled at, even through the wall of a container which would block air movement (McMonigle 2003). D. T. Anderson reports he can whistle at a certain pitch in the bug room and see all the big *Scolopendra* run around at the same time. (It is possible the deli containers have a strong acoustic vibration at a certain pitch.)

Centipedes often use the body to wrap around and immobilize prey. Banded *Scolopendra heros heros* x *Scolopendra heros castaneiceps* lab hybrid feeding on a grasshopper. © David T. Anderson

TRUNK SEGMENTS

The body, even of large, heavily armored individuals, feels pliable and soft, almost like the pliable body of a snake or worm. Most arthropods other than holometabolous larval insects have very rigid exoskeletons that crack audibly rather than bend when compressed. This also means a centipede that is fully hydrated can be much longer and fatter since the exoskeleton collapses and expands as the bulk within changes. Gravid and heavily fed females can look especially moribund. Inversely, sickly specimens that refuse to eat can become extremely short and thin over time. The same effect is carried over to dried specimens because they lose most of their width and length if not blown up with air or stuffed with cotton before drying. All the body plates are connected by stretchable membranes and even the tergites are not especially thick or hardened. The largest rotting specimens can disintegrate into nothing within weeks, unlike many other arthropods such as millipeds and beetles that can leave behind large pieces of their shells which persist for years.

This orange leg *Scolopendra dehaani* uses its strong, flexible body as a secondary appendage, twisting, flexing, constricting, and releasing as it tries to free itself from forceps.

The body is used as a secondary appendage. Centipedes can hold on with the back legs and lift the front to walk over large gaps or swing back from openings. (I am always curious how geophilomorphs ever end up in pitfall traps since they can drop most of the body down without losing their grip.) Many aggressive predators rear up the front of the body to pounce down on prey and some can even lift up the body to snatch prey from the air. The body can be swung around like a whip or wrapped around a prey item like a boa constrictor. It can be rapidly curled so the biting maxillipeds and pinching terminals can be brought to bear simultaneously against a threat. The body of a new mother can be humped gently over a brood of eggs or coiled tightly into a protective sphere.

The segments following the maxilliped bearing segment each contain one pair of legs, a lower plate known as the sternite, elastic pleurites on either side and usually a thickened upper plate called the tergite. This dorsal plate may be fused, reduced, enlarged, or split. The stretchable pleurites allow the centipede's

Some species or individuals can form a tight sphere with their body to wrap entirely around and conceal the eggs. *Scolopendra polymorpha*. © David T. Anderson

Sternites protect the bottom of the centipede body segments.

Dried specimens shrink up and collapse. *Scolopendra alternans*, Florida.

Pleurites are often light colored and hold the legs and spiracles. The flattened triangular spiracles are standard for the Family Scolopendridae, Subfamily Scolopendrinae.

In order for dried specimens to retain some resemblance to living animals, they require a process called "blowing," which is also used for soft bodied-caterpillars and certain beetle grubs. A hole is made in the rear and the guts are carefully removed (like a tube of tooth paste). A glass tube is inserted and the specimen is heated at low temperature until dry. The cavity can be stuffed with cotton following, or in place of, heating. © John Skierski

trunk segments to swell and shrink with food and water intake or loss. Pleurites also support and surround the legs and spiracles (usually). Unlike the pleural membrane of most insects and arachnids, it is usually easy to see the border between each centipede pleurite. In addition to pleurites on the sides, there are thin, intersegmental membranes that connect body segments to each other and allow very fat animals to expand greatly in length. These intersegmental membranes are why a dead animal or a molt can be stretched out much further than a living animal ever could, similar to the way a snake skin can be unnaturally stretched.

The terminal segment holds what are usually the second most highly adapted pair of

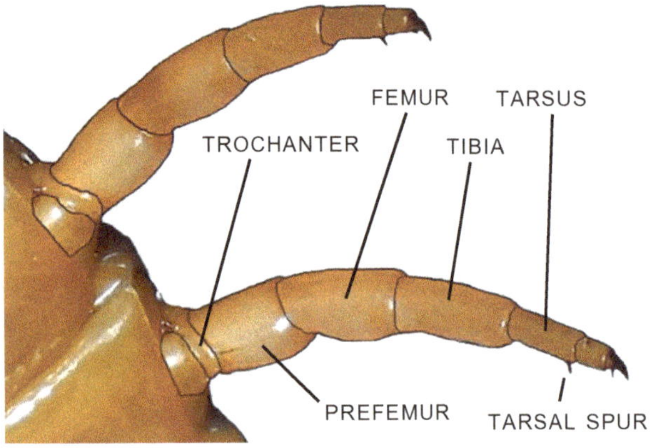

legs. This is also where the anus and the genital organs are located. On either side of the last sternite and anus is a plate known as the coxopleuron formed by the disjunct coxae of the terminal legs. The pleurites of the first (maxilliped) and last (terminal) segments are usually hardened or fused to other segments and there is little to no pleural region.

LEGS

Each leg is made up of the coxa, trochanter, prefemur, femur, and tarsus. The coxa and trochanter are usually very small while the prefemur and femur are usually the largest, widest segments. The tarsi can be very long in some lithobiomorphs and are hypersegmented in scutigeromorphs. The small segments making up the tarsus are known as tarsomeres. The very last tarsomere ends in one or two tarsal claws but additional spines can be located elsewhere on the tarsus and other leg segments.

The walking legs on each segment tend to be similar in size to the pair nearby, but often gradually enlarge along the entire length towards the rear. The last segment contains the terminal legs which are usually much larger and differently spined and shaped than the other legs. However, the second to last pair of

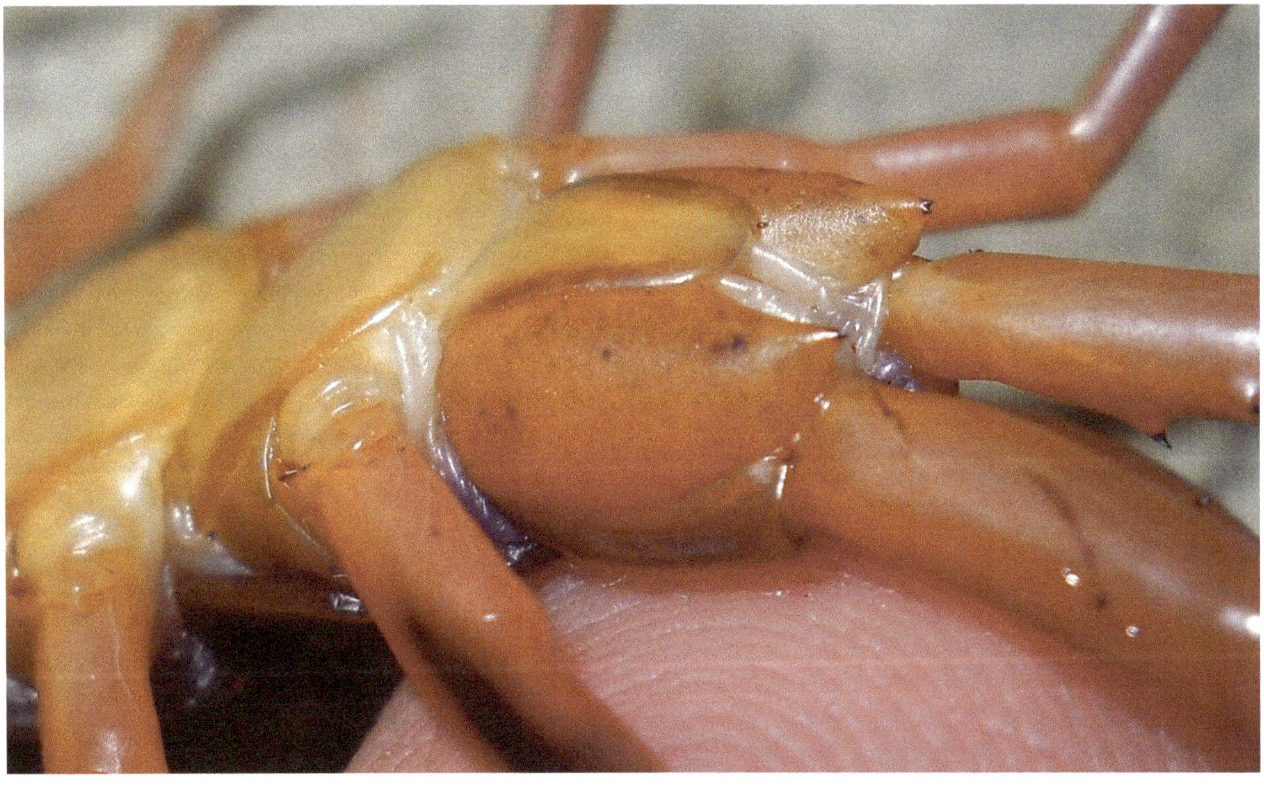

The rear coxopleura are greatly enlarged and the sternite is reduced. *Scolopendra dehaani* orange leg.

Terminals project from the rear of the last segment and coxopleuron. © John Skierski

legs can have some similarities and take over as a rear guard if the terminal legs are lost.

The hindmost pair or terminal legs (also known as "terminals") are usually very different from the other legs. Depending on the order, species, and level of development, the terminals will be widely variable in size, shape, and function. In scolopendromorphs the terminal prefemora and sometimes femora are thickened and covered in spines the other legs do not have, since they are used for prey capture like the rear pincer of the earwigs

Terminals are often brightly colored and can be shaken or reared up as warning flags. *Scolopendra heros arizonensis.* © John Skierski

(Dermaptera). There is often a large, inward facing spur near the end called the prefemoral process which has small spines at the end. Terminals can be thinner, longer, or even as highly segmented as the antennae, as in the Scutigeromorpha. In addition to being massive and pincer-like, they can be long, thick, brightly colored, and waved like warning flags. Some take this colorful warning a step further and give up any functionality in prey capture

Otostigmus scaber terminals are contrastingly colored for use as warning flags. © John Skierski

The most extreme adaptation is probably that seen in *Alipes* species whose terminals make noise when shaken in warning.

Scutigeromorpha take attenuate terminals to the extreme. In many species they are longer than the antennae.

to make warning noises like a rattlesnake with highly modified terminals. Since the rear legs are often used in courtship, they can be differently shaped or include small structures in males that aid in courtship recognition—this is most common in stone centipedes. The basal segments (coxa and trochanter) of these legs are likewise highly modified and appear to be part of the last body segment rather than part of the legs. Most legs project out from the sides of the body segment while the terminals face straight back.

Terminals can be pincer-like and aid in catching prey. *Scolopendra polymorpha*. © John Skierski

The number of legs a centipede has ranges widely but mostly according to the order, instar, or species, very rarely gender or geography. One piece of trivia often cited is no matter what the number, the quantity of leg pairs is always odd, but this is only true if certain facts are ignored. Every first instar Scutigeromorpha has four pairs while Lithobiomorpha have eight at second instar and twelve at third. (When they leave their mother, Craterostigmomorpha also have twelve legs.) Furthermore, the odd and even leg counts only hold true because the maxillipeds are counted as mouthparts when they are in fact legs. If these specialized legs were not ignored all mature centipedes would have an even number

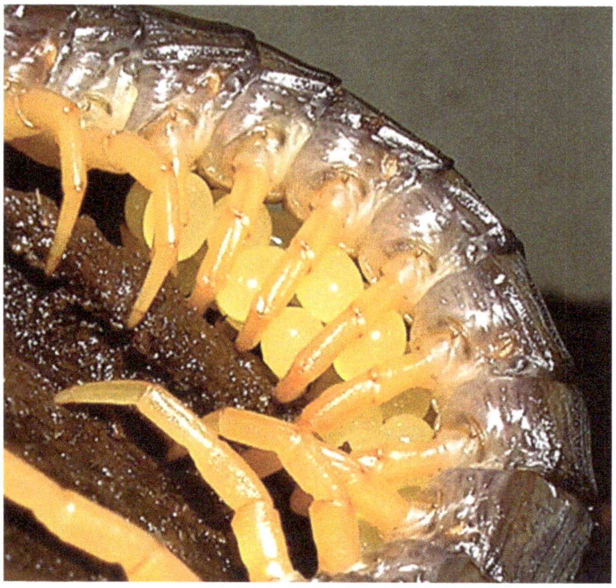

Spiracles are located on specific body segments. Roughly oval spiracles are standard for the Family Scolopendridae, Subfamily Otostigminae.

of pairs. Or if the terminals were likewise ignored whenever significantly specialized, many species would have an even number (the stridulatory terminals of *Alipes* spp., the highly attenuate terminals in house centipedes, etc. could just as easily be termed "non-legs" as maxillipeds are, since they are not used in locomotion). Lastly, there are always two legs per segment, so no matter how the number of pairs are counted, there are always an even number of legs.

RESPIRATORY SYSTEM

Most centipedes breathe through a system of tracheae similar to insects. The blood is not known to transport oxygen. The respiratory system begins with spiracles that open to the atmosphere. Spiracles are cuticle-lined openings located on each side of the body in the pleural region, usually every few body segments. The air enters a network of tubes that branch out so every cell of the body is connected. While it seems inefficient, massive centipedes and giant beetle grubs use this system and survive minimal ventilation that would suffocate a small mammal of half the weight. Little to no energy is used in breathing.

The Scutigeromorpha respiratory system is different from the other orders because the spiracles are located along the middle of the back, are not paired, and lead into a short vesicle before branching. Furthermore, a complex of tracheae surrounds the heart and the blood carries additional oxygen to the cells using the respiratory pigment hemocyanin. The modified respiratory system may allow for a higher metabolic rate.

The spectacular number of small air tubes (or trachea) that make up the tracheal system are often visible in specimens with light-colored tergites. © D. T. Anderson

EXOSKELETON AND GROWTH

The exoskeleton or body armor is composed of hardened plates connected by elastic integument. Much of the exoskeleton is composed of chitin, which does not grow once sclerotized, so the exoskeleton must be replaced as the animal inside grows. Young *Scolopendra* molt every month or two but as they grow larger, the molts slow down to two or three times a year. Smaller Scolopendromorphs and members of other orders experience shorter intervals between molts. Full-grown animals can spend over a year without molting. Adults do not need to grow but they do repair damage by molting.

Molting *Scolopendra hardwickei.* © D. T. Anderson

As mentioned earlier, there are two primary growth methods in centipedes. The first is anamorphic which means “changing form” and refers to the addition of body segments and legs in the early molts. Centipedes in the anamorphic group are called anamorphs (not to be confused with the 1990s toy *Animorphs* that transformed between human and animal). The second method, epimorphic, means “on form” and refers to species that keep the same number of legs and body segments after the post-embryo stage. For both epimorphs and anamorphs, most of the growth by volume is in the later instars, well after the full complement of legs has been reached.

A day to a few weeks before molting, the exoskeleton can become milky or discolored as the new and old exoskeletons begin to separate. Usually at this time the animal will stop feeding, often before any discoloration. Centipedes getting ready to molt may dig a scrape under decorations or make a hollow in the substrate. In nature, molting takes place in protected shelters but they are often exposed in captivity. Handling or feeding should not take place if an animal appears ready to molt.

When centipedes molt they extract themselves head first through the old shell like a

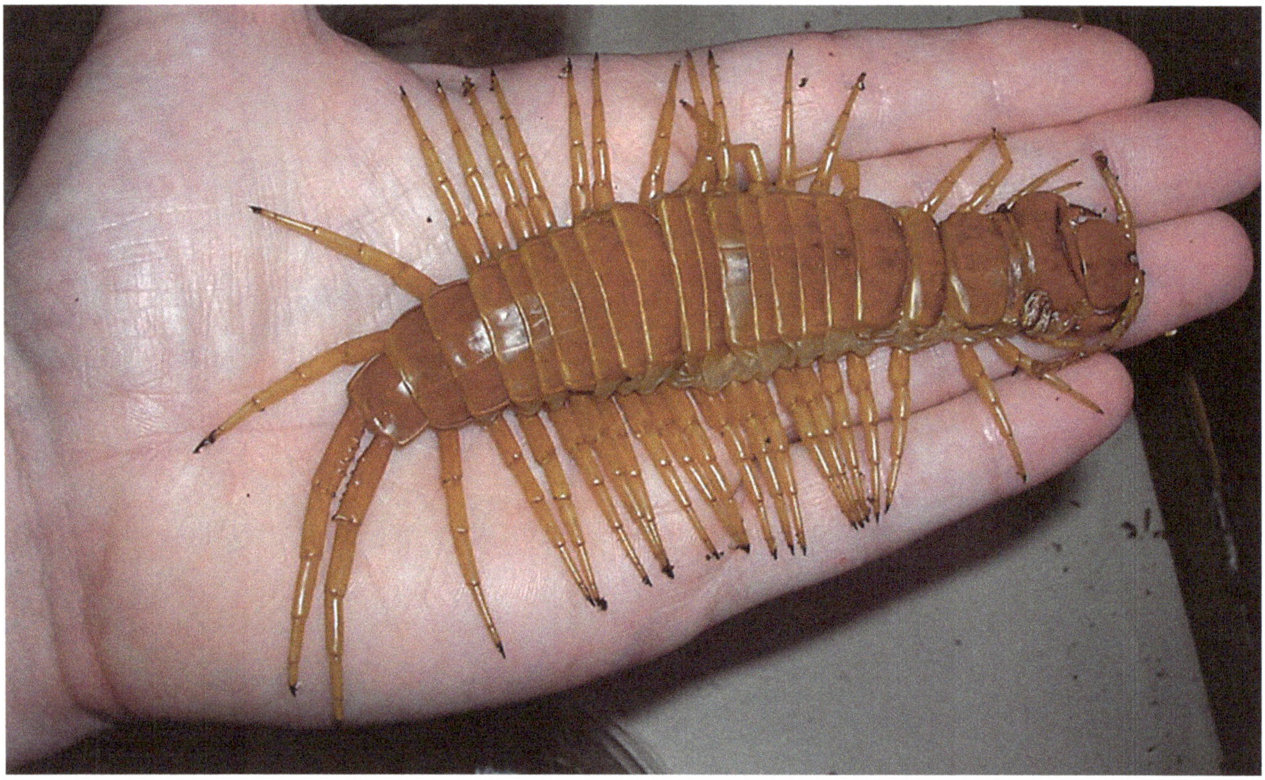

A molt. © David T. Anderson

Early instar *Scolopendra heros castaneiceps* molting. Hypopus mites move over before the shed is complete.

Young *Alipes grandidieri* molts and shortly after feeds on the shed exoskeleton.

caterpillar or pill milliped. The molt takes about half an hour (half as long for the first few molts). The new exoskeleton is a lighter color, soft, and nearly useless immediately after a molt. Uneaten, live prey left in the enclosure can take advantage of them at this time and eat the centipede's legs or chew holes through the body, killing it. Molting problems are very rare and are associated with low moisture or sick centipedes. Immediately after molting the new cuticle tends to look wrinkled and dented in a very unhealthy fashion, but this is normal. The exoskeleton hardens and returns to normal coloration and shape anywhere from half an hour up to several days depending on the species and the age of the specimen, since larger and thicker exoskeletons take longer. The centipede consumes its shed molt (except for the Scutigeromorpha) soon after the molt and before its new exoskeleton sclerotizes. It is obvious even when an adult has molted because

Immature *Scolopendra gigantea* molting.
© David T. Anderson

Rather small specimens or species tend to be white immediately after the molt. *Scolopendra alternans*, Haiti. © John Skierski

Pale coloration shortly after a molt. Adult *Scolopendra heros arizonensis*. © D. T. Anderson

the new armor is shinier and the color much brighter. If it is a young specimen, a notable size difference may be seen in the legs and antennae. If the hobbyist is wondering if a centipede has molted, it probably has not.

Centipedes are able to regenerate missing legs and antennae at each molt. Most if not all centipedes practice some degree of autotomy. That is, legs are designed to break off at a specific point, usually when grabbed by a predator (like certain other arthropod legs and lizard tails). In centipedes this point is between the trochanter and prefemur. Breakage elsewhere is rare. Closer to the body results in no regeneration, while breakage further out results in limited or defective regrowth. Unless the damage occurs within a few weeks of the next molt (Scutigeromorpha do not require more than a few weeks) the regenerated leg is almost the same size as the original legs in one molt. Similar rapid regeneration is seen in certain appendages of some other arthropods such as the amblypygid's antenniform leg or a mantid's walking leg. All legs of most other invertebrates capable of regrowing legs (crayfish, crab, spiders, walkingsticks, etc.) can take two, three,

Scutigera coleoptrata molting.

The color often becomes milky within a day or less of the impending molt. *Scolopendra subspinipes mutilans* from Jangsu province China. © Zhangjunduo

Normal coloration of *Scolopendra subspinipes mutilans* from Jangsu province China. © Zhangjunduo

Just after the molt, the body normally has a wrinkly, dented appearance. *Scolopendra gigantea*. © D. T. Anderson

The molt is consumed shortly after the body has been pulled from the old shell. *Scolopendra heros* subspecific lab hybrid. © David T. Anderson

or more molts before the regenerated appendage is large enough to be useful (McMonigle and Clausen 2002). Terminal legs regrow just as quickly and are large enough to be functional, but rarely ever return to full size in later molts (usually 70-90%). Regenerated terminal legs have a different spine arrangement, which can make identification difficult where spination is an important feature and both terminals have been regenerated. Like other legs, damaged maxillipeds regenerate, although the animal may not be able to eat enough to reach a molt. Regeneration in older centipedes is usually not as efficient, with some legs never reaching a normal size. Broken antennae regenerate, but not well, and they can still be missing segments after many molts. This regenerative handicap is likely because there is no specific joint designed for autotomy of the antennae.

The centipede exoskeleton does not retain moisture well. Even species that live in desert

This *Alipes grandidieri* was received missing 8 legs and all were regenerated in a single molt. The regenerated legs are often white in color following the molt, but they approach normal coloration after a few weeks.

Legs are rarely broken further out from the base, but when they are, regeneration does not work properly (post prefemoral segments are regenerated here in miniature). It is interesting these animals have an easier time replacing a whole leg than even a small tarsus. *Scolopendra dehaani.*

areas cannot survive extreme dryness. Desert species emerge from the earth following seasonal rains during the night to avoid the drying sun. If moisture is available in riparian areas, they can sometimes be found foraging during dry periods of the year and even during

Regenerated terminals are usually undersized and lack many spines. The angle of the photograph makes this regenerated terminal look smaller than it is. *Scolopendra polymorpha.* © John Skierski

the day though this is much less common. During dry, hot periods they dig deep in the soil and use the earth and stones as their shield against desiccation. *Scolopendra* can be kept in a dry cage as long as there is a water dish, but they need to constantly replace lost fluids. This may be stressful to large animals and is deadly to small ones. It is important to provide consistent moisture in captivity. If a centipede is placed in a well-vented, empty cage without a water dish, small species or young will take on potato chip-like qualities in less than a day, while larger animals may survive as long as two days (though irreparable damage occurs the first day). The ubiquitous house centipede can survive dry conditions a little longer than others, but not well.

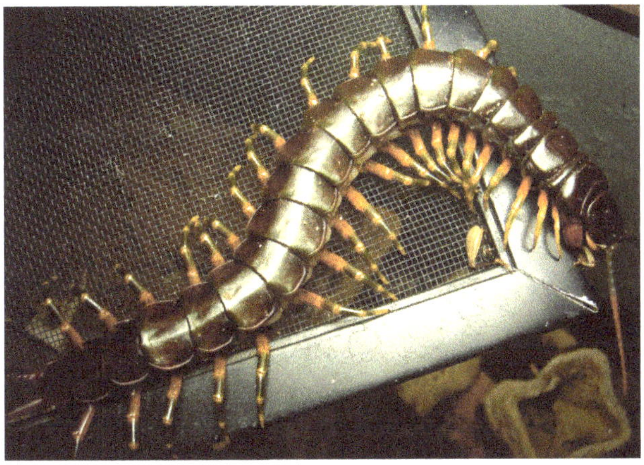

The dark-colored *Scolopendra gigantea* changes to an orange color a day or two before the molt. © David T. Anderson

These second instar *Scolopendra subspinipes mutilans* color up with quite a different pattern than their mother. © Kelly Swift

Following a few molts the juvenile coloration of *Scolopendra subspinipes mutilans* is still very different from the adult. © Zhangjunduo

COLORATION

Warning coloration is common for large species with painful bites. Many are striped in orange and black or yellow and black. Banding can also be cryptic in nature since banded and unbanded forms of some species are found according to habitat. Banding and spotting on different scutigerid centipedes does not contrast bright colors and likely breaks up the outline against mottled bark and lichen-covered stones. Some *Scolopendra heros* have a black rear end thought to be mistaken for the head (Stoops and Martin 1995). Also, the antenna and terminal legs are both orange on these, while the legs are yellow, enhancing the head-to-rear similarity. Certainly there is no advantage to having one's genitals mutilated instead of the head, but the terminal legs are very large and protect this end. A small number of *S. heros* seen in the wild will be missing one or both of these legs (or have one disproportionately regenerated) since they were likely shed

Scolopendra heros heros banded, from the Big Bend area of Texas, USA. © David T. Anderson

Scolopendra heros castaneiceps from central Texas, USA. © David T. Anderson

Captive-produced hybrid of banded *S. h. heros* and *S. h. castaneiceps*. © David T. Anderson

It is easy to see why some believe *Scolopendra heros heros* color is used to confuse the head and rear. © John Skierski

in defense. Having the head and rear look similar may work to discourage predation altogether rather than encouraging a predator to attack the back end. A predator thinking about eating a centipede probably wants to know which end is the head (I know I would) and if it cannot tell it may look elsewhere for its next meal. *Scolopendra polymorpha* also often have this kind of coloration (though not so highly developed) with both ends marked in orange and the central body greenish tan. Nevertheless, there are quite a few centipedes whose heads are brightly colored in contrast to the rest of the body, including very small specimens that cannot be advertising a painful bite.

Sometimes the exoskeleton changes color over a molt or series of molts. An initial color change is normal for nearly all species because they transform from white or yellow first instars to the pigmented second instars early in development. Most gradually darken or lighten in color as they approach adulthood while they

This banded *Scolopendra heros arizonensis* was only fed only crickets from an early instar and it changed from the normal coloration, almost losing the dark markings entirely (they are more noticeable in the photo than in person), while the antennae and termals turned brownish.

Standard banded *Scolopendra heros arizonensis* coloration seen in wild and captive specimens.

maintain the same general pattern. In some species the second instars and small immatures are a completely different color than adults and this only changes over many molts. *Scolopendra subspinipes morsitans* and certain *Scolopendra subspinipes* have distinctly different juvenile and adult color patterns. The least common expression of color change is seen when a centipede develops a different color over many molts due to some aspect of the captive habitat (whether food, humidity, or something else). This has only been documented in *Scolopendra heros arizonensis* (McMonigle 2003). Though rare, it is probably not isolated to a single specimen of a single subspecies.

LONGEVITY

The chilopod total lifespan is similar to long-lived pet arthropods like assassin bugs, rhinoceros cockroaches, or whipscorpions, but short compared to certain common pet tarantulas that live for decades. There are no distinct signs of adulthood; maturity is assumed when the centipede reaches nominal adult size. Unlike insects, adults continue to molt after maturity. Males that have bred and females that have produced eggs continue to molt. House and stone centipedes can be reared to adulthood in less than a year under captive conditions, but can take two years outdoors in temperate climates. Smaller *Scolopendra* including *S. viridis* can reach adulthood in under a year while large species like *S. heros* and *S. subspinipes* usually take three years to mature. As with virtually all predatory arthropods, the time taken to reach adulthood is greatly affected by the quality and amount of food and temperature. Poor care and sub-starvation can stretch out development to five years and extremely terrible care might stretch the time out further, but more likely would result in premature death rather than longer life. Following maturity, the adults continue to live two to five years. If a *Scolopendra heros* is purchased at 2nd instar, it may be alive six or seven years later. Reports of total longevity in excess of seven years are not only very rare, but tend to lack basic documentation. Anything less than three years from hatchling to death, for almost any species, would only be the result of poor husbandry, the bottom of a boot, can of pesticide, etc.

FUTURE

Centipedes as a whole are at little risk of being extirpated from global habitats and communities though species with limited habitats such as a single cave or cave system and those found in tropical rainforests or on small islands are at risk of habitat destruction from human activity. The Serpent Island centipede, *Scolopendra abnormis*, from the island of Mauritius, the former home of the dodo bird, is the only centipede currently listed as threatened. This is of course due to the intense habitat repurposement by humans. It is curious how many unknown species have been lost with the removal of the rainforests, but we will never know. Also, the global spread of *Lithobius forficatus* and *Scutigera coleoptrata* with European colonists hundreds of years ago may have affected some competing species. Our native *Scolopendra heros*, mostly the subspecies *S. h. castaneiceps*, has been exterminated from many areas in Texas as shopping malls and housing developments have taken up large areas of their habitat. Also, it is a species people love to hate and smashed animals are a common site where fisherman and hikers frequent. Still it is a common and widespread animal, so hopefully the only concern would be the isolation of some populations. Humans have certainly been sold millions of dollars in pesticides for the express purpose of murdering the "scary" but helpful and harmless house centipede. Myths about the danger of large species certainly bring about the worldwide smashing of incredible numbers of *Scolopendra* daily. With man's intense hatred for centipedes it is amazing only *S. abnormis* is threatened. Unfortunately, it is unlikely the centipede averse would buy this book in hopes of learning a greater appreciation and respect for these helpful predators.

Tiger Legs molting. © David T. Anderson

TOXICOLOGY AND HANDLING

Centipede bites are not considered life threatening since there is not one shred of irrefutable evidence linking a centipede bite directly to the death of a human. Venom can have different effects on different people. Direct handling or holding of any centipede is not recommended. Handling can severely damage or kill the animal.

"For humans a bite is painful but not dangerous" (Werner and Olson 1994).

"Like tarantulas, this vast lack of medical reports on them more than likely indicates they are not medically significant" (Breene 1997).

"There are no deadly species of centipedes and their bites are needlessly feared" (Evans 2007).

There are two common legends of death caused by centipedes. The first is an army officer (or cowboy) bitten on the chest or back (remarkably by every leg on the centipede's body) a hundred or so years ago. Baerg (1924) relates this as a legend concerning a Confederate officer who died two days after one such monster walked across his chest leaving a trail of venomous wounds with every footstep as it walked. The second yarn concerns a very small, very young girl in a third-world county bitten on the soft spot of her head (whether the centipede flew to her head or was attempting to swallow her whole is unknown) over half a century ago. She has been transformed into "small children" (Dunn 1993) or a baby from the Philippines (Greenaway 2000), but most often is a 7-year-old Filipino that died twenty-nine hours after the bite. Cloudsley-Thompson (1958) relays this "true" story but offers no citation or reference. Even if these tall tales were true, a person is about as likely to die from being hit on the head by a meteor and thousands of times more likely to die from falling out of bed.

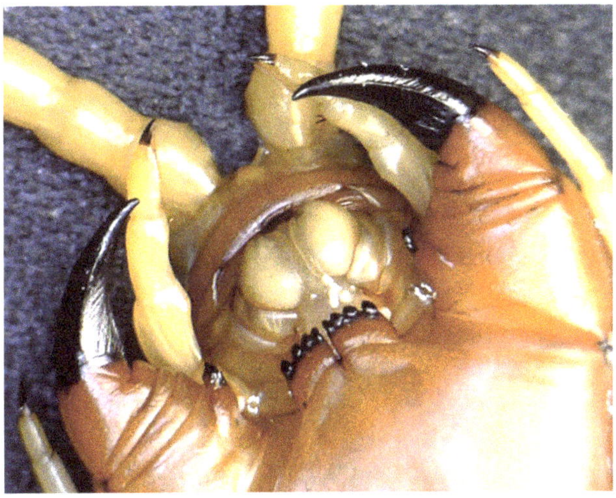

The venom-introducing maxillipeds of the largest centipedes certainly look foreboding. Golden *Scolopendra polymorpha* from southwestern Texas, normally mislabeled *Scolopendra heros*. © David T. Anderson

The only documented death related to a centipede bite was confirmed as a death due to a bacterial infection. The fatality followed after a Turkish man returned to the hospital four days after coming in with a centipede bite, this

Handling large *Scolopendra* seems to work for some people, but the animal may be accidentally damaged or killed. *Scolopendra heros castaneiceps.*

time with necrotizing fasciitis, a bacterial infection, believed to have led to multiple organ failure. Necrotizing fasciitis usually occurs in people who have advanced alcoholism, HIV, chronic renal failure, etc., and can be caused by a minor scratch or injection of insulin (Serinken et al. 2004). Cause of death = bacterial infection.

The puncture wound from the bite of a large centipede can be just as dangerous as a puncture from a thumbtack or dirty nail. The wound should be washed and cleaned and antiseptic applied. Consult a physician regarding tetanus shots. As with all wounds, signs of secondary infection should never be ignored. Centipedes are venomous, so invertebrate dealers do not sell them to anyone under the age of 18 for liability reasons (a doctor's visit is expensive even if it is simply for tetanus shots).

"Centipedes seem to exert a weird fascination on the morbid appetites of the hysterical and insane" (Cloudsley-Thompson 1958). Humans love to make up stories about centipede bites. Many bite reports are highly dubious if not altogether fictitious, especially when it comes to the smaller species. In the case of the largest *Scolopendra*, various reports (Lewis 1981, McMonigle 2003) claim the pain is similar in intensity to a bee sting and subsides primarily within half an hour, even if swelling and minor irritation last longer. However, there are some victims who compare the pain to that of unset broken bones or fingernails being removed with pliers (both of which they likely never experienced). It is always possible a bite could hit just the right nerve, such as under the fingernail, but dramatization seems a more likely culprit. Unfortunately a centipede bite seems to be a badge of courage and the longer one say it hurts, the more oohs and aahs acquired. Yet there are no accolades to be acquired for a bacterial infection. It may be difficult to discern fact from intricate fable. However detailed they may seem, fables always lack basic documentation.

"One common belief in New Mexico is that centipedes burrow into your skin. I never heard of this myth before moving here, but I get asked about it several times a week since opening our exhibit" (Wade Harrell pers. comm.). This could be based on urban legend or related to one report of a centipede found in a hole burrowed into a cadaver's forearm (Harada et al. 1999).

"I was told by one of my Filipino field assistants who was bitten in the foot that it is extremely painful and he could walk normally only after about a month. No doctor was consulted, however, so it is possible that he would have been less affected with proper treatment" (C. Schwarz pers. comm.).

The chance of getting bitten by most smaller centipedes such as *Scutigera* is quite rare—the vast majority of bite stories are fiction.

Scolopendromorpha bites from large specimens can be extremely painful. Normal indications include localized pain and inflammation. Other, possibly psychosomatic, symptoms can include nausea, numbness, dizziness, headaches, heart palpitations, and vomiting. Numerous anecdotal reports are available which detail pain lasting from a few seconds to half a day. Lewis (1981) reports the bite of *Scolopendra morsitans* from Nigeria produces a sharp, short-lived pain, whereas bite reports from 1870 for the same species from a different locality included vomiting, headache, and swelling, followed later by a black lesion. While it is possible there are differences in venom toxicity for different populations of the same species, or reactions for different people, the black skin area suggests poor hygiene and the result of bacterial infection.

There are no commercially available centipede antivenins because the venom is not life threatening. Centipede bite information is anecdotal and the few rumors of more severe reactions cannot be traced to their roots. No commercial research (except for use as a pain blocker or cancer cell inhibitor) is done on the subject, which belies the insignificance of centipede venom. In contrast, extensive research has been, and continues to be done on venomous reptiles and certain arachnids like the widow spiders because their venoms are medically significant. Admittedly, there are probably more deaths from choking on hotdogs in a year than by all venomous animals combined. According to the CDC, about five people die a year from venomous snakebites in the United States and nobody has died from a widow spider bite in more than a decade. However, complications from snakebites can continue for years and availability of antivenins may be responsible for low fatality rates in the case of these reptiles.

Only a single case of anaphylactic shock has ever been recorded for a centipede bite. The size and identity of the creature responsible for the bite were neither confirmed nor directly measured. A 23-year-old man in Thailand was bitten on his left little finger by a 9" (23 cm) specimen (this reported size alone is almost certainly an exaggeration). The man was treated and released (Supakthanasiri et al. 2004). The authors explain that the venom contains seratonin (5-hydroxytryptamine) in the South American *Scolopendra viridicornis* or cytolysin in the North American *Scolopendra heros*, and there is an occurrence of histamine release by these venoms, but the centipede itself was not available for testing and the man did not return after going home from the hospital.

PERSONAL EXPERIENCES WITH ENVENOMATION

When my younger daughter was between the ages of five and seven she would always say, when asked, that centipedes were her favorite bug because she liked to hold them. I would say, "Are you sure you don't mean millipeds are your favorite?" I would show her a milliped and say, "These are the ones you mean, right?" Her answer was no, she really meant centipedes. She liked to go out and lift up rocks to play with the stone centipedes outside. (In northern Ohio the largest common garden centipedes are 30 mm *Lithobius forficatus*.) These centipedes are soft, easy to see, and do not smell when handled like the common garden millipeds. The strangest part about this conversation is it happened more than once.

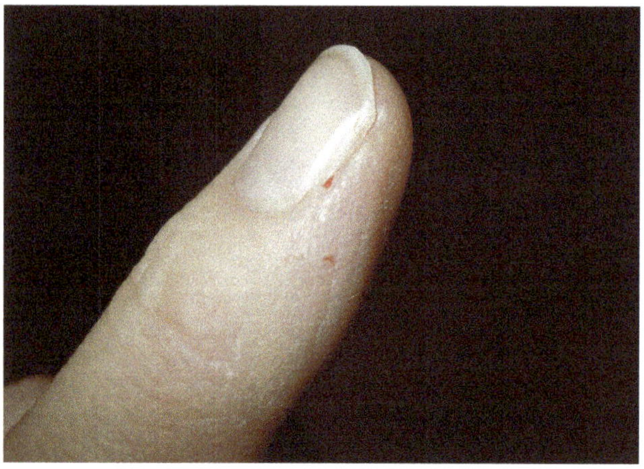

Maxilliped punctures from an adult Floridian *Scolopendra alternans*.

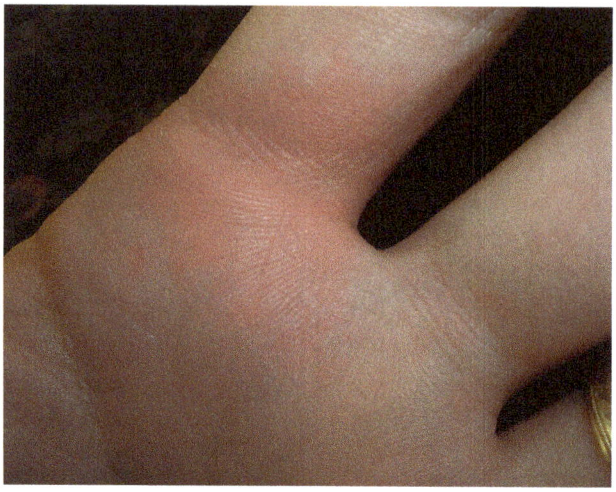

A few minutes after the bite, red splotches can be seen on the skin near the bite. (Punctures are in the soft skin between the fingers.) The pain lasted about thirty minutes, 2:04 pm, 6/6/2007, adult *Scolopendra heros castaneiceps*.

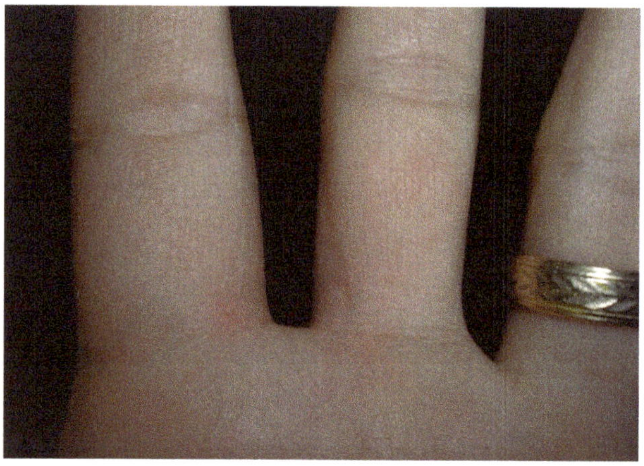

The pain subsided as swelling began which lasted into the evening, 5:43 pm, 6/6/2007, adult *Scolopendra heros castaneiceps*.

Of course, she never played with a large *Scolopendra*. Some people I have watched pick up monstrous *Scolopendra heros* without getting bitten, but they do not try to handle the really aggressive individuals or other species. They also make sure the animal keeps moving. I am always amazed when they are not bitten. I have handled countless *Scutigera* and *Lithobius* because they generally cannot inflict pain or damage. (I do not place them in my

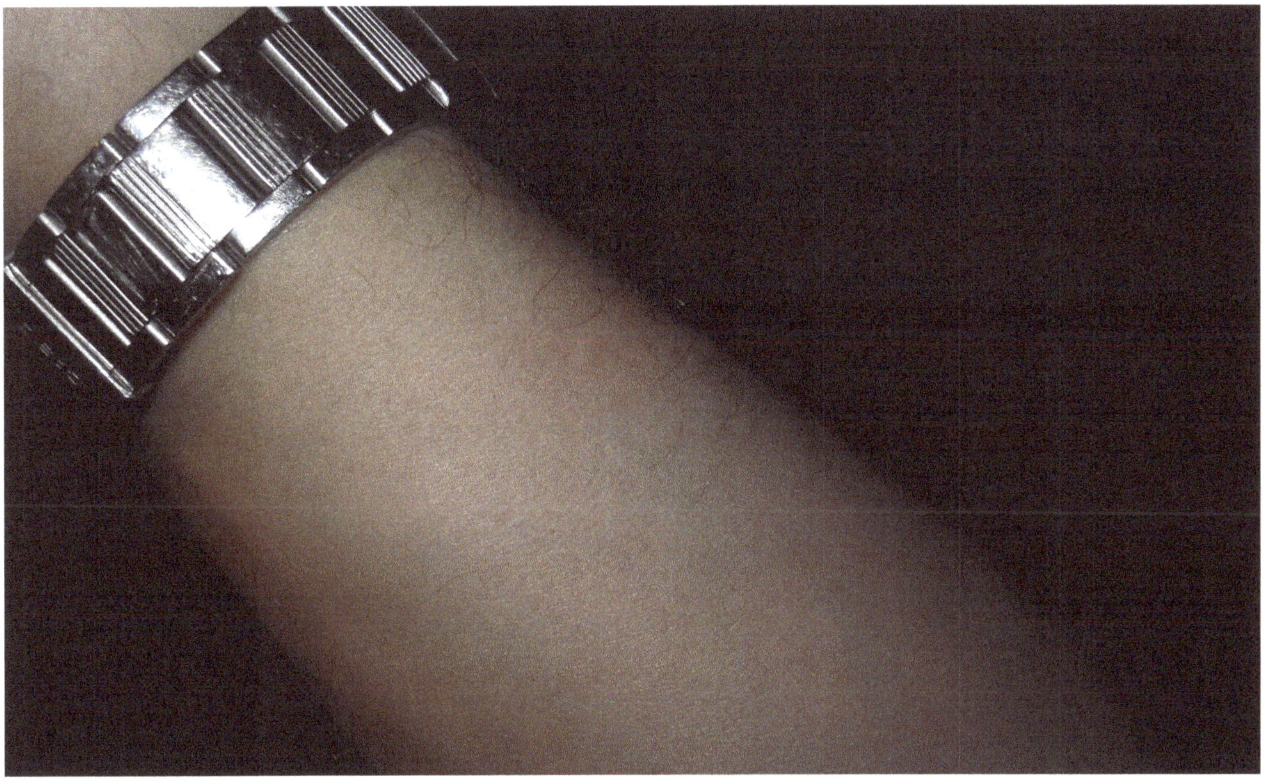

Smaller *Cryptops* cannot bite the skin of the fingers or hand, but this little guy picked a thin spot of skin on the inner arm. The pain lasted for 20-30 seconds and then the tiny raised welt seen here lasted about half an hour.

armpit.) I have also purposefully held large and small *Scolopendra polymorpha* and *Scolopendra subspinipes mutilans* and have not yet been bitten by these species. This is because they are less apt to bite, but moreso because handling has been an extremely rare event. I have only handled large adult *S. heros* twice and never on purpose.

The most painful bite I have experienced was also my favorite. It was during an attempt to rehouse a large adult *S. heros castaneiceps* when somehow it ended up near my shoulder facing downward. I suppressed the immediate instinct to fling my arm and whip the huge centipede across the room. (I did not want to risk hurting it.) Instead I thought, "I'll put my hand on the floor, let it walk down and place a cup over it." I was initially quite proud of my cool reaction and quick thinking. The centipede walked down my upper arm, down my forearm, and across my wrist to the base of my fingers. Only at this time did it carefully bite into the thin skin between my fingers, sinking a maxilliped deeply into either side of the most tender flesh. It bit patiently and injected as much venom as it possibly could in the softest and most painful possible area. Only after watching in slow motion horror and feeling the maxillipeds sink in deeply did I finally flick it across the room (and then put a cup over it). If I had not suppressed my first instinct with "quick thinking" I probably would not have been bitten. The bite hurt worse than any previous or since, but it hurt for half an hour and then was swollen for a few hours (the pain subsided when the swelling began). This episode was my favorite because I realized I cannot outsmart a centipede.

One of the most feared species is the large *Scolopendra subspinipes* since its bites are considered very painful, can result in intense swelling, and are the subject of much centipede bite lore. I once fed a cricket to a large Chinese *Scolopendra subspinipes* (in a shallow tub, not a proper cage) and it bit me in the finger. There were 1 mm lacerations and a small amount of blood where the maxillipeds had sunken in. I was very worried for the first twenty minutes waiting for my arm to swell up or to suddenly feel excruciating waves of pain, but it was just the standard tingling pain that was mostly gone after half an hour. It was certainly uncomfortable and as painful as a yellow jacket sting, but a whimper compared to stories I had heard. It was comparable to a bite on the finger I received from a large adult Florida *Scolopendra alternans*. Had I expected something less painful I might have been impressed.

Only once did I purposefully put myself in a position where a bite was inevitable. My biggest adult female *Scolopendra mirabilis* jumped its cage during maintenance (again, too short of tub). I knew the only way I could be certain to prevent damage or loss of the animal was to grab it, there was no time to hunt down capture tools. Some pain was worth it to keep this beauty alive and healthy. I held it firmly as it sunk its fangs into my skin and pumped venom to its heart's content. Sometimes one does not mind taking one for the team if the team is pretty enough.

With various bites there have been small cuts where the claws broke the surface. Still, the initial mechanical damage does not usually cause more pain than any cut of similar size and depth. There is generally a few seconds to a minute before any sort of pain kicks in. For me, the pain ranges in intensity between that of a sweat bee sting and that of a yellow jacket. It is strange compared to other venomous arthropod bites because the pain goes in and out in waves. The discomfort fades in and out and after the first few minutes, if I concentrate on something else, it is almost forgotten. With bites from large animals, the pain lasts for fifteen to thirty minutes, at which point I have to poke the bite area to be sure it still hurts. If minor discomfort from pushing on the tiny wounds is included, the time period may be greatly enhanced. Minor, localized swelling, when present, lasts from half an hour up to five hours. As mentioned, the onset of swelling usually signifies the termination of any sharp or rhythmic pain.

HANDLING

Scutigeromorpha, Lithobiomorpha, Geophilomorpha, and Scolopendromorpha (under two-inch body size only!) are usually not strong enough to break through normal human skin, but be extremely careful if you are a cricket or a human with tissue paper skin. The largest stone centipedes do open and close their maxillipeds, trying to bite, which can be very unnerving since the attempt can be felt. One scientist who wanted to test bite reactions would place large stone centipedes in his

Dendrothereua sp., Texas. © John Skierski

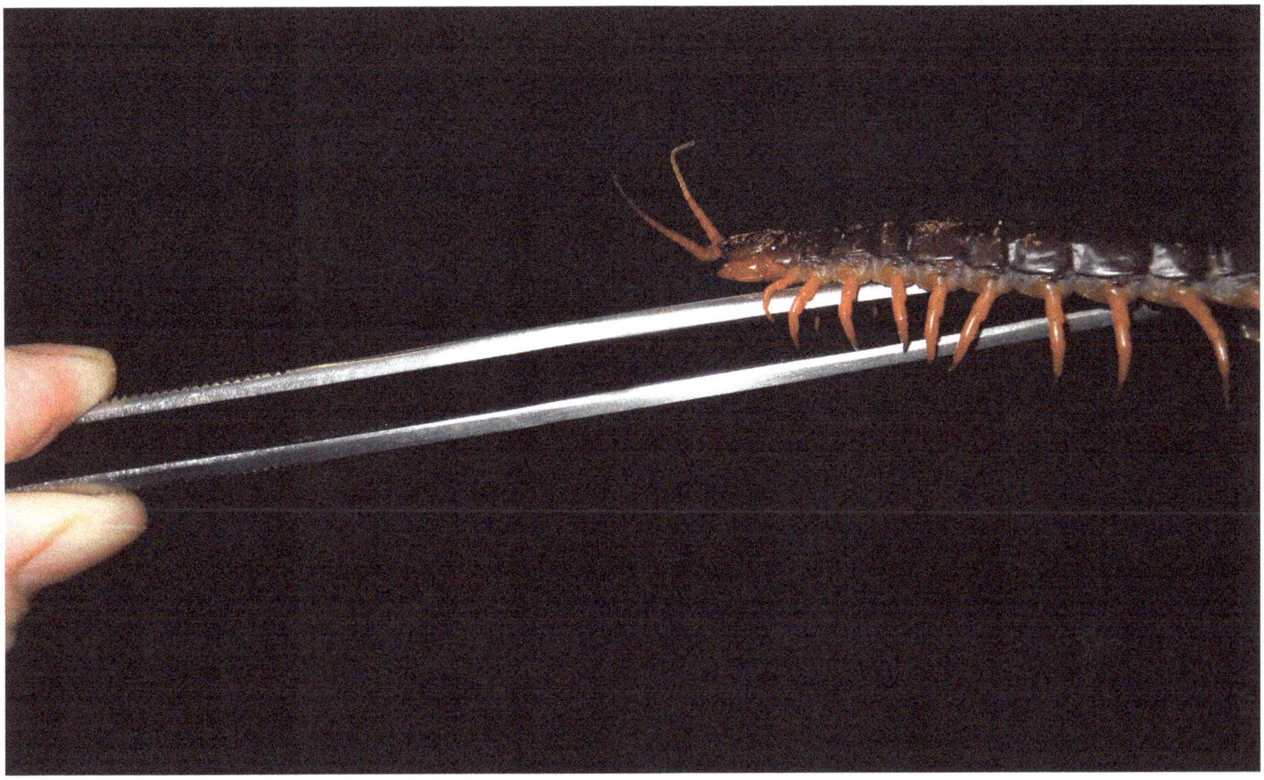

Specimens will try to come back up the forceps even when restrained tightly. *Scolopendra dehaani* orange leg. This seemed a lot closer when the picture was taken.

armpit where the skin is very thin, as the only way to obtain a reaction (Hase 1928).

Centipedes are not angry monsters as they are commonly depicted. Large crickets frighten all but the largest centipedes. These secretive animals do not attack people and prefer to run and hide. Of course they will bite if grabbed or stepped on (or pushed headfirst into the armpit).

While venomous vertebrates such as cobras can have more potent venom when young, invertebrates are different. The venom glands grow larger and more productive with each molt. When rearing young, it is helpful to keep in mind that a 3″ (7.5 cm) *S. s. subspinipes* may hardly impress with its bite but the same animal if it grows to 8″ (~20 cm) is likely to make a lasting impression. Hobbyists sometimes become less wary over time but caution should always be used.

Proper handling methods always keep the centipede's venomous maxillipeds far from the hobbyist's skin. It is difficult to entirely avoid handling since cages need to be cleaned occasionally. When it is necessary to transfer or remove the centipede from its container, this should take place over a tall plastic tub, five-gallon bucket, or even the bathtub as a precaution. If the bucket is half filled with water, it cushions the fall and slows down the centipede (centipedes that do not sink are rather impressive serpentine swimmers). The whole container (or containers, when transferring to a new cage) can be set down inside a large storage bin.

A common handling method involves containers and a sheet of cardboard. Cardboard is placed over the top as lid is removed, another container with the same size mouth is placed

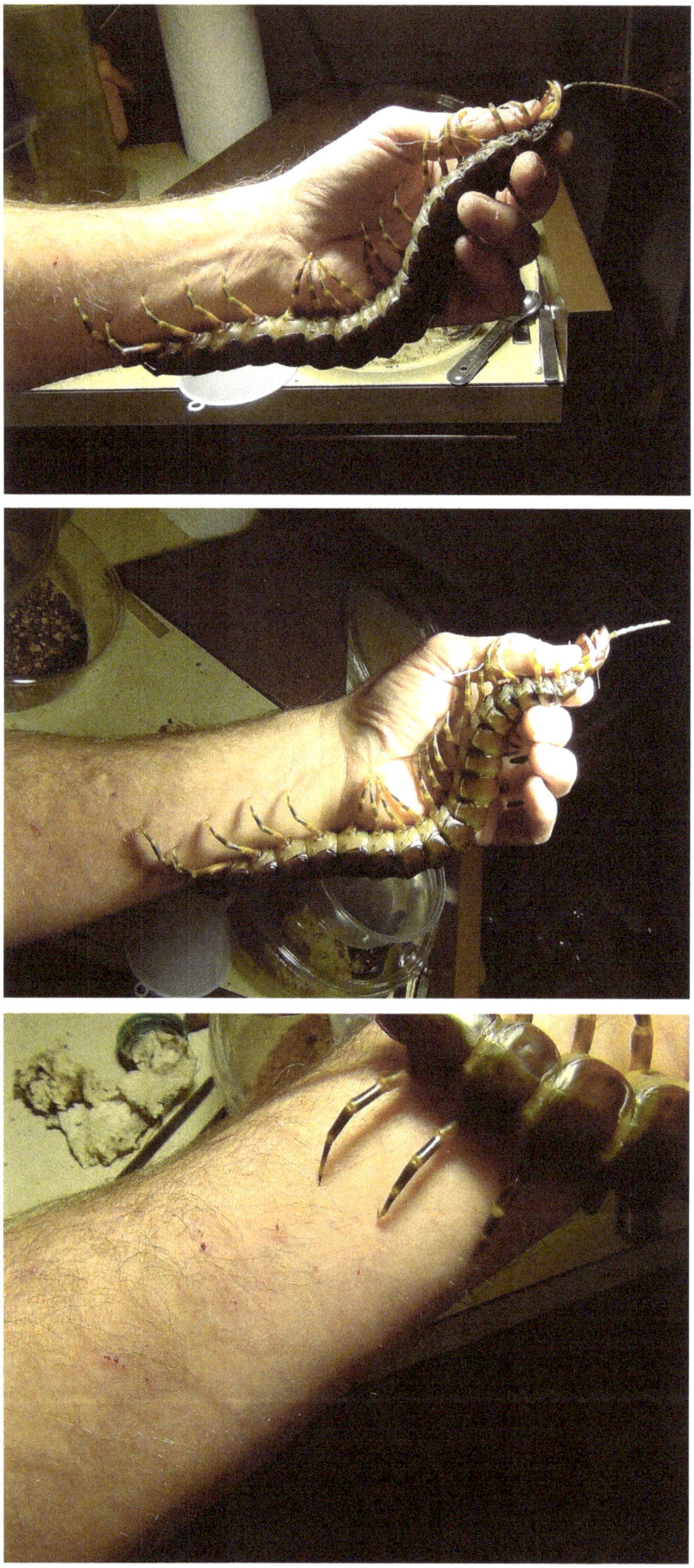

Puncture wounds are not dangerous themselves, though bacterial infection is always a possibility. © David T. Anderson

Unrestrained *Scolopendra gigantea* climbing the forceps. This female was picked up to show the recently accepted spermatophore that is still visible. © David T. Anderson

against other side of the cardboard. The cardboard is removed and both containers are tilted sideways to allow the animal to walk through, the cardboard is replaced when the animal is safely on the other side, and then the lid is replaced. If a centipede ends up on the floor, it is covered with another container and a piece of cardboard slid underneath.

Forceps can hold tightly onto the flexible body without harm. (The black scars are because this is a recent import and it arrived damaged.)

A set of long forceps can be used to move centipedes. Even if the container method is used, it is good practice to have a pair of forceps handy (an extra pair or a dowel rod is needed as well, since many animals refuse to let go of the forceps and climb rapidly towards your hand). Never grasp by the legs or antennae because they tear off. The body of the centipede must be gripped tightly across the body

An anesthetized centipede. © David T. Anderson

segments. It will violently try to wiggle away and stretch its body in every direction to grab objects. A large animal's attempts to bite the forceps can be felt as vibrations in the keeper's hand and may be a little unsettling. Patient use of forceps does not harm large-bodied scolopendromorphs. Of course it is possible to puncture or tear an animal in half, but the keeper would have to be terribly careless. Some hobbyists use hooks meant to move snakes. While rather amusing, use of snake hooks is begging for trouble. Unlike snakes, centipedes have legs.

"I have used a 1" (2.5 cm) diameter round artist paintbrush with a long handle to move centipedes a short distance, from one cage to another while both cages are side by side inside a big plastic tub. The centipede attacks the brush like prey when touched, chomps on the bristles and wraps its body around it. While so engaged, I lift the centipede into the other container by moving the brush over, and when it figures out the brush is not edible prey (it just takes a few seconds) it releases. It does not work every time so a backup plan is a good idea. The centipede figures it out pretty fast, so I always do this while the cages are in a big tub with smooth sides to be safe. Also, sometimes I have to drop the brush in a hurry in case it decides to climb the handle! Not exactly a foolproof method, but if done inside a tub and with backup wrangling tools (forceps) nearby it is a quick way to move them with no risk to the animal and minimal risk to the handler" (Wade Harrell pers. comm.).

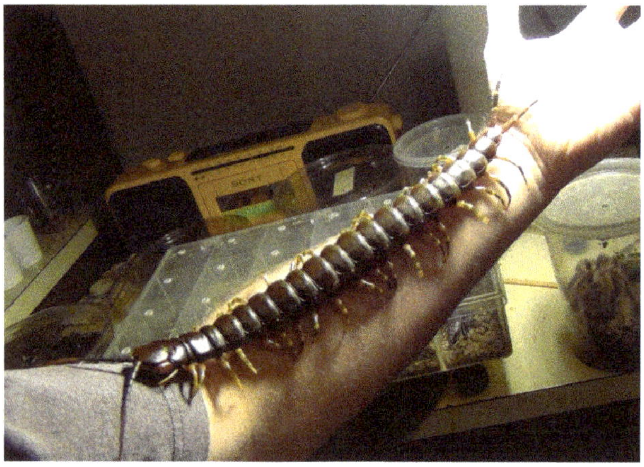

This centipede is about ten inches long. © David T. Anderson

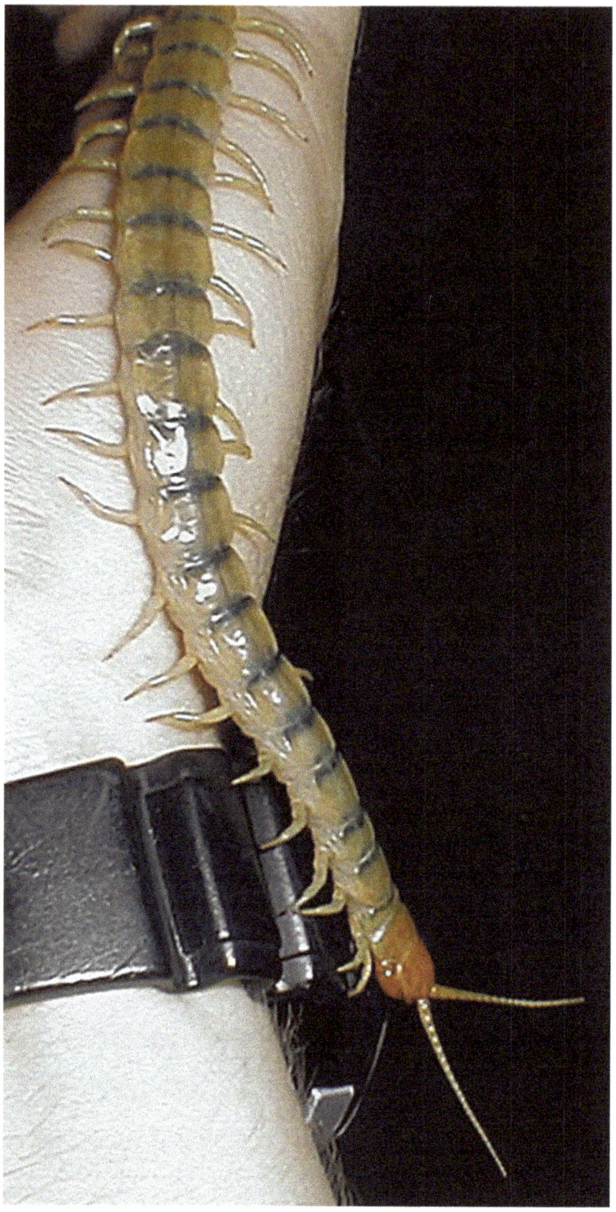

An active *Scolopendra polymorpha* adult.

Some people do hold large scolopendromorphs in their hands for photos. Most of the pictures in this text are a trick. The animals had just molted and at this time their exoskeleton is generally too soft for them to attempt a bite. There is still a risk, especially as the exoskeleton approaches normal coloration and rigidity. In addition to the risk to the handler, posting these photos on the Internet may also give the impression to new keepers or even children that handling centipedes is a good idea, which is irresponsible.

HABITATS

Centipedes have conquered nearly every land area on earth, including most islands, but they are not found in the frozen expanse of Antarctica. And yet the Antarctic's largest animal construct, a mobile, manned research station completed in 2013, British Antarctic Survey's Halley VI station is nicknamed the "blue centipede." Representatives of the Class Chilopoda are found across all climates where they inhabit forests, plains, caves, deserts, mountains, beaches, tide pools, and the homes of other animals from termites to humans. They are among the most prevalent predators in woodland settings and can be found beneath most rocks and logs. The majority never leave the safety of the ground though a few inhabit the walls of caves and homes, and a decent number climb trees. Arboreal representatives are chiefly from the Scutigeromorpha though at least one of the Lithobiomorpha (*Lithobius variegatus* from the UK) is regularly found more than ten feet above the ground (Nichols et al. 1971). *Scolopendra* from rainforests also regularly climb at night. "I have seen more specimens of this genus in trees than on the ground" (C. Schwarz pers. comm.). Even some of the soil-dwelling Geophilomorpha are found among debris under the loose bark of trees and can be surprisingly far above ground. The rarest species are restricted to limited, non-littoral habitats on islands, or within cave systems. Nearly every large-bodied species is found in the tropics, even those whose range extends into warmer temperate areas. However, the greatest diversity may be found in temperate climates among the myriad tiny species, many of which may be unknown to science. There

"Trinidad: This is a small species, but as you can see from the picture, what it lacks in size, it makes up in coloration! I collected a few of the offspring but was unable to raise them" (Martin Gamache pers. comm.). © Martin Gamache

Panay Island, Philippines, rainforest habitat. © Christian Schwarz

This red *Scolopendra* sp. was caught in a pitfall trap and placed in a bucket, Panay. © Christian Schwarz

Large *Scolopendra subcrustalis* climbing a tree in the rainforest, Panay. © Christian Schwarz

Large *Scolopendra subcrustalis* climbing a tree in the rainforest, Panay. © Christian Schwarz

Juvenile *Scolopendra* sp. out at night in the rainforest. Panay. © Christian Schwarz

Scutigerid hanging out in a cave, Panay (two of the right legs are autotomized). © Christian Schwarz

Scutigerid on a branch, Panay (three rear legs are autotomized). © Christian Schwarz

Scolopendra viridicornis habitat in Para, Suriname. (Specimen photograph in species chapter.) © Martin Gamache

could possibly be more tiny species from the tropics, but the largest single family, the Lithobiidae, with nearly a third of all known species, is mostly temperate. Wherever centipedes live they must have access to moisture and yet they can thrive in desert areas simply by hiding away underground until the ground is softened by seasonal monsoons. Amazonian species must deal with the opposite problem of seasonally flooded forests where they must either climb or stay submerged, a condition the seashore-loving Geophilomorpha conquered long ago. Centipedes produce large numbers of young, often a hundred or more over the life of one female. Although they do not produce thousands of eggs like many marine crustaceans and some insects and spiders, the vast majority must likewise never make it to maturity in nature. Usually, the goal of artificial habitats is to transform a natural 98% die off to an artificial 98% survival by offering conditions that lack dangerous extremes and providing all requirements needed for survival. Each beginning keeper can still face some trial and error, but reviewing the following pages which detail habitat experiences should prevent unnecessary error. A lot of mistakes may only seem foolish in hindsight.

ARTIFICIAL ENCLOSURES

Proper caging is the most important aspect of keeping centipedes. An improper cage will shorten the centipede's life. Worst of all, a bad cage can make a hobbyist quit.

Some enthusiasts have left the hobby because of an escaped animal. While the mandibles and first maxillae of the largest centipede cannot chew through much of anything, they are escape artists and will locate and squeeze through gaps a keeper may not see or register as large enough. They can climb edges and wires to reach gaps around the lid. Simplicity in caging is safest because it is difficult to block all possible escape routes in complex glass and screen terraria. Many hobbyists keep their pets in the readily available, plastic, preformed cages or plastic tubs which can be somewhat safe if the keeper is 100% perfect at snapping the lid on correctly, but these cages are not escape proof. A ten-inch (25 cm) centipede is a beautiful creature in its cage. The same animal running around the house can unnerve even the most ardent of keepers. If the spouse happens to glimpse or find out about the escaped critter, the centipedes or spouse may be gone the next day. At least one hobbyist no longer keeps centipedes because the escapee bit his girlfriend. He promised her he would never keep them again.

The recommended cage for large centipedes is a plastic jar with screw-top lid. There is no latch or pop-on lid to worry about snapping correctly. The cage does not shatter and the lid does not pop off when the cage falls to the floor. The strongest centipede cannot screw the lid off or chew through the side. Large plastic jars (square or round) are available at most department stores for a few dollars. The sides of most

Chaco Forest in Formosa, Argentina. *Scolopendra* sp. "Funny story, my two friends shouted at me that they found some frogs (they know I love frogs). By the time I get to the piece of rubbish where they found the frogs, they were searching somewhere else. I lifted the piece of cardboard and there was this giant centipede. I shouted at them, "Very funny guys, there is just a huge centipede." They came back running to see it. They had only seen the frogs in a different layer of the cardboard. It pays to bring several people on these trips! Here is a shot of the animal itself and another beside one of the amphibians. (There were two amphibian species, including frogs, this one was a toad.)" (Martin Gamache pers. comm.) © Martin Gamache

Scolopendra heros castaneiceps and *Scolopendra polymorpha* habitat in New Mexico, USA.
© John Lorbiecke

Scutigeromorpha species. French Guiana under a rock formation.
© Martin Gamache

This handsome centipede from the Subfamily Otostigminae was unearthed in French Guiana along with these termites, *Nasutitermes* sp.
© Martin Gamache

Jar with a screw-top lid.

Jar lid vents. These vents would be far too large if the ambient humidity were below 20%.

are see-through and excellent for viewing. Plastic jars are available in any size but the most common are half-gallon and one-gallon (~2-4 liters), which are perfect for most centipedes. Large cages are nice but the animals do not require a cage as wide as their length. However, a container that is taller than the centipede's length can help prevent escapes as they do not climb smooth plastic well unless there are holes or textures that allow them to get a foothold. It is important the lid is the correct size for the container and screws on securely. Three to twelve holes can be drilled or melted in the in the lid for ventilation. The holes should be 3/16" (0.5 cm) in diameter or smaller. Thumbtack holes at the lowest edge of the container (below the substrate) are useful to drain off accidental overwatering.

There is likely to always be some gap between recommended and actual caging, especially when there are large numbers involved. When rearing up sixty *Scolopendra heros*, among others, there is often not enough time to eat sixty gallons of pretzels before they reach a decent size. Also, if rearing a large number, it takes much less time to feed individually when each container has a snap on lid. In this situation tall deli containers (32 oz) may be advisable as the centipede will not be able to scale the sides.

What`s this? It is the letter awaiting hobbyists who do not secure their centipede cages.

If the hobbyist did not choose to use a screw-top cage the question may then become how to reacquire the animal. Homes tend to be very dry and if a centipede is placed in an empty bowl the animal will dehydrate within hours. However, large desert species especially are adept at shielding their bodies from desiccation by burrowing under objects and staying in tight cracks. There is no way to effectively bait centipedes, but damp towels placed on the floor around the house might attract the drying

Communal caging is not always a good idea even for less aggressive species. *Alipes grandidieri* and conspecific meal.

Large specimens can do well with dry vented caging and a water bowl. This is a rather large *Scolopendra polymorpha.* (The small, dark pellets are actually 7 mm *Dynastes tityus* frass.)

This was a dry setup with water bowl at the Smithsonian insect zoo in Washington, DC. *Scolopendra heros arizonensis.*

animal to shelter. Even if, against all odds, an escapee is reacquired in seemingly perfect shape, it is likely to stop eating or wither away within a few months. Prevention is key.

Humidity and moisture are extremely important and often confused. Dehydrated animals cannot drink humidity. It is entirely possible for a centipede to dry out when moisture is limited even if the humidity is 100%. In a cage with limited ventilation it is possible for condensation to deposit water high on the cage walls and desiccate the substrate over time, especially if there are minor temperature variations on different sides of the cage (common for many shelves or side lighting). Extremely low ambient humidity and a well-vented cage cause substrate to dry out too fast for the keeper to keep up easily and one mistake will spell death. Excessive ventilation is never harmful in itself, it just requires constant replenishment of moisture. It is like tilting back in a chair, as the keeper is setting himself or herself up for a fall. The further the leaning, the more terrible and certain the fall. Even "desert" centipedes will not survive without moisture. Never place an incandescent bulb directly over any part of a centipede's cage. Even the most gigantic centipede will sustain massive damage within a few days as the extremities will dry out. Moist substrate and minimal ventilation conserve moisture. A water dish is not necessary. Ventilation is important so centipedes can breathe, but the total combined area of vent holes does not ever need to be more than a square quarter-inch (8 x 8 mm). Extremely wet conditions can indirectly harm centipedes. Wet, rotting prey pollutes the air in the cage. Another drawback is that high humidity along with overfeeding can lead to mite infestations. Mites feed on half-eaten prey

Plastic plants can add some color to the enclosure without requiring lighting and care. *Scolopendra polymorpha.*

and thrive with high moisture. Mites dry out easily, but not as easily as centipedes.

Some hobbyists keep centipedes on dry substrate with the well-vented lid of a molded plastic creature keeper. A shallow water dish must always be available and the container's lid should be partly covered if ambient humidity is low. This type of setup can reduce the negative effects of overfeeding and is acceptable for large display animals, but it will prevent reproduction and can kill smaller centipedes. These commonly available cages snap tightly but it is not always easy to see if the trap door was fully snapped down (or if the small plastic nubs wore out). One way to guard against escape and desiccation is to place these cages within very large plastic totes.

Centipedes cannot climb smooth surfaces. The legs end in one or two simple claws, they do not have specialized feet like many insects and arachnids that adhere to smooth, vertical surfaces. Even house centipedes with comparatively complex tarsi are found stranded in the bathtub from time to time. Although unable to climb smooth surfaces, many prop themselves up on their hind legs and reach the top of any container not taller than the combined length of the body, rear legs, and highest decoration in the cage. They cannot climb the sides of an all-glass aquarium but may climb the silicone bead running up the corners. (The silicone bead can be carefully removed with a sharp razorblade though the tank may never hold water.)

Live plants in a large, complex enclosure. © John Skierski

Display tank decorations give the centipede a place to hide or offer a handsome-looking habitat. Rocks are probably the most common natural shelter for centipedes in nature, but in captivity they can result in smashed or dismembered animals (rock plus centipede often equals death when wild collecting as well). A flat piece of slate with four rock feet, hot-glued to the underside as spacers, can make an acceptable hide. The feet should be in contact with the bottom to prevent collapse. Commonly available and useful hides are slabs of cork bark or shallow coconut huts. Even the scutigeromorpha are thigmotactic and prefer to rest with object touching the sides of their bodies, so greatly arched hides are often ignored. Plants do well in centipede cages since they usually are not dug up or chewed on, but the only purpose they serve is aesthetics. Live plants may help keep the air quality higher in moist,

Caging with rock caverns. © David T. Anderson

Terrarium with locking lid. © David T. Anderson

limited-ventilation cages but they consume rather than produce oxygen after the lights go out. Pothos vine (*Epipremnum aureum*) is the most practical since it grows easily from cuttings and ambient room light is usually enough to keep it alive and healthy.

Common substrates include moist sand, potting soil, peat, coconut fiber, vermiculite, gravel, and wood shavings (avoid substrates from the pine family). Preventing soggy or bone-dry conditions is more important than the type of substrate used. Vermiculite and most sandless soils adhere to the centipede's exoskeleton and mask the beauty of the animal. Gravel or wood chip substrates usually prevent digging behavior. A shallow layer of aquarium gravel placed beneath the primary substrate is useful to drain off excess water unless the species being housed is one that will thoroughly mix the gravel into the substrate. Potting soil or coconut fiber mixed with 25% sand makes a great substrate. The ratio is not magic—75% coarse sand and 25% coco fiber can work too, but the dampness of the substrate is usually difficult to see when comprised mostly of sand. The animals seem to enjoy digging in and out of the mixture. The sand improves support, drainage and helps to clean the centipede's body as it burrows. If the animal burrows constantly, sand scrapes attached mites and dirt off flat areas of the armor.

With proper care, substrate replacement is generally only done when animals are moved to new cages, or if the build-up of prey material leads to pest infestations or sickly animals. Replacement following a problem is helpful,

Group of 32 oz deli-cups commonly used for rearing medium-sized juveniles. © David T. Anderson

but afterwards is often too late and usually too little. Substrate replacement will need to be done at regular intervals if husbandry is not corrected and it will stress animals, especially young ones. With correct feeding frequency, centipedes can be kept on the same substrate for seven years or more. Clean-up crews can be

Success does not always require large size. 16 oz deli-cup with female and brood, *Scolopendra heros castaneiceps*. © David T. Anderson

32 oz deli-cups. © David T. Anderson

useful to consume leftovers and frass. Springtails are better suited than isopods for many applications since large centipedes cannot eat them.

THE ARGUMENT FOR PET DIRT

Substrate depth is a debate as old as centipede keeping itself because the correct answer is simply not one or the other. It is important to understand what different depths offer to choose that which best fits the size or stage of the animal, as well as the keeper's care style. Substrate can be as deep as desired, but a depth approximately twice the width of the centipede's body is plenty. Most breeders employ shallow substrate to keep an eye on the animals. Specimens of a variety of species have been reared on shallow substrate from dispersal to maturity with high survival rates. Viewing allows for an improved level of care and feeding, especially when the animals are very small and require a higher level of attention. Inasmuch as shallow substrate may allow a new keeper to overfeed more easily or allow the substrate to dry out and desiccate the animal, greater depth can be preferable. However, deep substrate can also lead to starvation, massive pest infestations, and prey revenge.

Large tubs are popular cages. © David T. Anderson

Though deeper substrate maintains moisture far better, the keeper is more likely to forget to water an "empty" cage. If the diligent keeper eventually forgets about the animal they have not seen in six months, desiccation through apathy is no less possible than with shallow substrate. Furthermore, moisture deep in the substrate is often not usable because centipedes will often attempt to flee (run around the surface looking for escape) rather than dig to find moisture. For rearing immatures I find shallow substrate offers the best results for practically any centipede, but I have also chosen to use deep substrate for breeding, brooding, and cohabitating certain species. Deep substrate can make it easier not to bother the female and doom her brood as females who are disturbed during brooding often eat eggs or offspring, but if she is never seen it is entirely

This female *Scolopendra heros castaneiceps* lived a long, healthy life in a plastic jar cage.

Large tubs are popular cages. These are the permanent homes of captive-hatched and reared *Scolopendra heros heros* and *Scolopendra gigantea*. © David T. Anderson

possible the mother could end up eating most or all the babies before any are viewed by the keeper. There is also the danger of digging up a deep-burrowed female and dooming the eggs or larvae anyway, since checks must eventually be done.

TEMPERATURE

Temperature is not a major consideration in the normal care of centipedes because room temperature 72°-75° F (22°-24° C) works well for nearly all species. Most have little trouble with temperatures from 65°-90° F (18-32° C). Some of the worm-like Geophilomorpha die at temperatures above 80° F (27° C). Desert species can survive down to 40° F for short periods and even tropicals rarely die in shipping from a little cold. Temperate species can survive much lower temperatures and many stone centipedes still grow and reproduce in the mid-sixties. Seasonal temperature shifts have not proven necessary to breed any of the many species that have so far been kept in captivity. Keep in mind that temperature is not a major consideration assuming they are kept at normal room temperatures. Temperate species may be able to survive in an unheated garage in the colder areas, but even the more resilient species can be cooked in the summer or frozen in the winter.

Heat pad on enclosure wall. © David T. Anderson

Although centipedes are found in every climate, the giants like *Scolopendra* are found primarily in the tropics and extend into the warmer temperate regions. In North America, *Scolopendra* species are found across the southern U.S. and none reach Canada. Even in these colder temperate areas, the temperature a few feet below ground is a consistent temperature around 55° F (13° C), even when the outside temperature is below zero. This suggests they either do not dig deeply enough to overwinter or are unable to predict the coming of winter.

Maximum growth and feeding potential can be reached by keeping animals in the high seventies to low eighties (26°-29° C). Some of the *Scolopendra* species especially will grow more quickly if kept warmer. A heat mat or tape can be placed on a wall of the cage just above the substrate. Many species will be found "basking" near this heat source. The vast majority of centipede species will be just as healthy without the added heat but can take much longer to mature. The food quality and quantity is much more important to growth, but only inasmuch as temperature does not impede feeding. Lastly, if your home temperature is below 72° F (<22° C), added heat may be necessary to keep certain tropicals from barely feeding or growing. Huge temperature shifts are not needed—keeping the cages on a top shelf rather than a bottom shelf may be enough.

Scolopendra gigantea enjoying the warmth of an artificial heat source. © David T. Anderson

Otostigminae species, *in situ* in Western Australia habitat. Female guarding eggs. (CC-SA) Wendy Eiby

THERE ARE NO VIRUSES or bacteria commonly seen to harm captive centipedes. Certainly wild animals may be afflicted with microscopic ills as well as visible protozoans, barely visible nematodes, tiny parasitic wasps and flies, or gigantic horse hair worms. However, these are incredibly rare in captivity. Those problems observed with any regularity are listed in the following paragraphs. Low long-term survival of imported specimens might be blamed on invisible ills, but poor transit conditions are the more likely culprit. Wild-caught *Scolopendra alternans*, *Scolopendra heros,* and *Scolopendra polymorpha* that have not been subjected to the poor conditions and long handling times of import, almost always live full lives in minimally adequate conditions. Most of the time, imported specimens display evidence in the way of damage, lethargy, or lack of appetite when first acquired and still drag on for months or years. Mortally wounded centipedes can die very slowly. Still, many imports look perfectly fine at first but then quickly turn for the worse.

Mites are the most commonly observed problem in captivity. Parasitic mites are not seen but they would be rather large and look like a polyp attached between the segments. The commonly seen problem is the hypopus (or phoretic) stage of free living mites. This is a special stage found only in certain types of mites, most commonly the ubiquitous grain mite *Acarus siro*, that attaches to the edges or surface of hardened plates of larger arthropods for transportation purposes. Another commonly seen species is a pink colored hypopus stage mite seen on wild *S. polymorpha* from California. The hypopus mites look like a crust and do not seem to move (often confused as "eggs") but under magnification it is easy to see they have legs and move actively. The mites are not only are capable of telling plastic from an insect exoskeleton, but also can tell living from dead transportation. Objects of similar surface texture placed in the cage will not attract the mites. If the centipede molts they will move

Hypopus stage mites in low numbers are harmless. Here a few are seen around the base of the maxilliped and between tergites 6-8.

These red hypopus stage mites are sometimes seen on wild *Scolopendra polymorpha* from California.

This coating is just dirt, not mites.

over to the freshly molted animal even if the exoskeleton is uneaten. If it dies the mites abandon ship, but they will return if the centipede begins to rot and becomes a large meal.

Minor mite infestations are common and do not harm the centipede. Wild-caught centipedes can already have some on them. Small infestations may even be slightly helpful in alerting the keeper to improper conditions. A few dozen barely noticeable mites growing into a mess of hundreds indicates improper feeding. At this time the cage should be cleaned, substrate discarded, clean-up crew added, and the feeding schedule reduced in quantity, volume, or prey type. Large infestations which go untreated result in thick coatings of thousands of mites. In extreme cases they are harmful and can lead to partial blockage of the spiracles and death of the centipede. Still, under such rancid conditions it would be difficult to prove chemical byproducts of decomposing prey were not the more significant health hazards.

Wild-caught specimens often look healthy from above but the large-bodied Asian imports are especially prone to arriving with body wall damage. This is not visible in the opaque-sided cups they are often sold in and can be difficult to see in the clear tubs unless they climb the sides. The damage looks like a massive crack across or between the sternites filled in with dried black blood. The surrounding area is often bruised and black. This may heal at the next molt but usually such specimens stop eating after a few months and then die a few weeks later. It is not rare for more than half the animals in a recently imported group to have body wall damage. Such massive trauma is almost certainly related to collection methods (maybe the collectors step on the animals with thick boots to catch them).

"Leg rot" is a less common health issue primarily restricted to wild-caught animals. The

"Leg rot" damage of unknown origin (the claws are much smaller than the black areas). © John Skierski

condition is rarely noticed when the animal is first acquired and then later assumed to have been caused by captive conditions when finally noticed. Most often, damage to the leg extremities is a result of physical damage or extremity desiccation, but the damage can grow worse over time or be healed at the next molt. Moisture is often blamed but I have kept various species extremely damp for years and have never seen leg rot spontaneously develop. Most

Dead leg areas of a dying animal. © John Skierski

importantly, the majority of centipedes require very high humidity to stay alive. Wild humidity fluctuations that cause the extremities to dry out and die, other creatures in the cage that chew on the legs like nematodes, or substrate full of rotting prey remains infested with fungus might lead to leg rot problem in captivity.

Body wall damage, scarring, and dried blood are pretty common on wild-caught centipedes from Asia.

Small *Scolopendra heros arizonensis* missing legs, collection damage.

Internal damage developed over a period of a few weeks on a wild-caught *Scolopendra dehaani* orange leg. Internal damage was accompanied by advanced lethargy at twenty days following a sixty-second drowning event. Similar discoloration developed on a specimen from the same import, but without lethargy.

"When this centipede was halfway finished molting, a cricket crawled on its back and started eating it. The centipede would have died if I had not seen it waving the front of its body back and forth in an unusual way, and quickly removed the cricket. This shows the animal after it molted from the injury, it took two more molts for it to look normal—I still have this centipede. I was amazed it survived" (Anderson pers. comm.). © David T. Anderson.

Current picture (Feb.2014) of the *Scolopendra heros heros* specimen damaged by a cricket (June 2011). Even after a number of molts and a few years this specimen is still imperfect. © David T. Anderson

Lack of circulation in an old or dying animal is another possibility. "Leg rot" may simply be a symptom of senescence. A similar looking problem develops in certain millipeds and praying mantids when specimens become very old and yet drag on. This should not be confused with physical damage caused by closing the ends of the legs under a lid or by *Zophobas morio* mealworms or *Pynoscelus surinamensis* cockroaches that can bite the centipede's legs and should never be used as prey.

Nematodes may be disproportionately blamed for health issues but evidence is lacking since they have not been observed to attack living animals. Nematodes do scavenge on dead, rotting centipedes and are all but invisible when buried in the substrate. Living stone centipedes and worm centipedes feed on nematodes, not the other way around. According to its describer, nematodes seem to be the preferred food of *Lithobius muticus* (C.L. Koch, 1844). Nematodes are among the most common organisms on earth and are present in every centipede enclosure unless great pains have been taken to heat treat decorations and every speck of introduced soil including any stuck to the feet of the centipede upon placement in the cage and on every prey item ever offered. These tiny, clear worms are about as thick as a human hair and a fraction of an inch (1-3 mm) long. Nematodes may reproduce to plague proportions on leftover prey materials and in the case of extreme infestations might be a cause for "leg rot." Tens of thousands to hundreds of thousands of starving, biting, near-microscopic nematodes may be able to wear out areas of the exocuticle over many months.

Other than nematodes, phorid flies, mites, and other scavengers are sometimes mistaken for causing death, but the presence of a massive, rotting food source will allow them to reproduce in numbers. The reproductive speed and prowess of phorids is especially impressive since they can turn a dead centipede into a writhing mass of maggots within days. Phorid flies can smell a good carcass and are small enough to fly in through window screen from

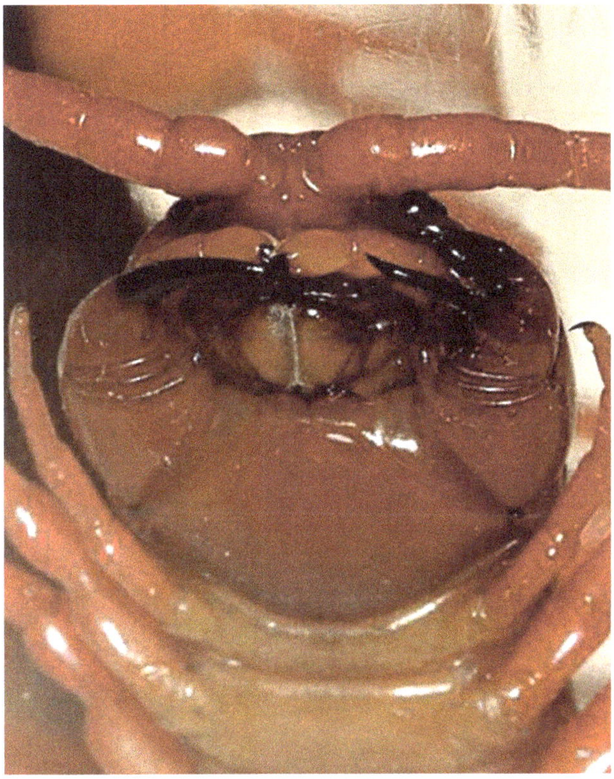

One maxilliped with three claws on a wild-caught specimen. © Martin Gamache

outdoors, though they are commonly associated indoors with prey cultures, especially crickets. Nematodes and grain mites could take weeks to fully consume a dead centipede but there may already be massive quantities of them in the enclosure that had been feeding on prey leftovers. These small invertebrates are advantageous scavengers rather than centipede killers.

A rare problem in captivity is seen with phorid flies that normally only eat dead arthropods. Larvae are sometimes able to attack centipedes and enter through a spiracle. It is uncommon, only reported for a few Otostigminae (from the Scolopendromorpha), and tends to only occur at one spiracle. Maggots may be seen entering and exiting the spiracle and the side of the tergite swells up. The phorids do not generally kill the centipede but the damaged area does not heal even when the animal molts. Prevention consists of removing uneaten insect carcasses to prevent maggots in the cage. Every once in a while a wild-caught specimen arrives with a damaged area that appears similar to damage caused by phorids in captivity.

If a number of legs or the front or back section of the body seem to be paralyzed there is nothing that can be done. This is usually considered a sign of physical trauma or partial desiccation but pesticide exposure could display similar results. Paralyzed sections are very rare if the animal has not died after a few days. Some centipedes seem to slow down overall (not falling on their back or losing full use of any appendages) and if pampered for a few months they may recover. Lethargy, like paralyzed sections, may be confused with dying, but in these unusual cases the animal can live on for weeks or months.

Tergal segmentation defect. *Scolopendra gigantea*, captive-bred. © David T. Anderson

Malfunctioning autotomy is seen from time to time. Sometimes scolopendromorphs that regenerate a terminal leg after a molt will lose it again for no apparent reason a few weeks later. I have seen this happen as many as six times in a row in the same individual. Improved diet might help but there is probably an internal defect in the regenerated limb. Also, if a cage decoration is moved and accidentally smashes the middle of the body (not enough to kill the animal), it is not rare for a terminal to fall off. Autotomy is highly advanced in the scutigeromorphs but when they are dying the legs fall off one by one. Unless the problem is noticed very early on it is often too late. Causes include feeding with only fruit flies for too long or waiting too long between feedings. How long is too long? A few days to a month or more depending on the age of the animal, species, and temperature.

Lastly, we come to the rare developmental or genetic defects. Certainly these could be caused by heat shock during embryo development but it would still encompass a small percent of the genetically predisposed. The most common is a disruption in development of the segments. This same problem is also seen from time to time in millipeds (McMonigle 2012a) and somewhat regularly for beetle larvae (McMonigle 2012b), but also cockroaches, pseudoscorpions, and likely any segmented arthropod. Segment defects are harmless unless associated with severe internal malformation. The appearance may change with growth but this type of defect can never heal. Afflicted animals should not be used for breeding. Other defects are far more rare and include divided legs with multiple claws (including maxillipeds) and body wall indentations or pleural malformations.

Scolopendra morsitans with mites. © John Skierski

CENTIPEDE FOOD

ALTHOUGH OFTEN CONSIDERED predatory as a whole, only the Scutigeromorpha are true obligatory predators—they will not scavenge or feed on vegetable matter (notwithstanding dehydrated animals drinking sugary fluids). They can catch small prey using just the articulate maxillipeds, but employ additional front leg pairs to keep larger prey from escaping. Other centipedes can grab onto prey with any of the legs, including the terminals. Then they wrap their strong, flexible body and legs around the food to immobilize it for the bite. The Scolopendromorpha are primarily predators while the Lithobiomorpha and Geophilomorpha can make vegetable matter a large part of their diet even when heavily outfitted for predation (i.e. large maxillipeds) (see Lithobiomorpha chapter). Centipedes as a whole can eat nearly any terrestrial invertebrate from the nearly microscopic to the very large, including a range of smaller vertebrates. The largest of the monstrous Scolopendromorpha have been observed feeding on small vertebrates, including pinkie mice, small lizards, amphibians, bats, and baby birds. Feeding on small rodents is likely an artifact of captivity though a number of authors cite the ability of large species to eat mice (Bucherl 1971, Stoops and Martin 1995, Breene 1997, Sandefer 1998b, Bruins 1999, Greenaway 2000, Ditzel 2002, O'Toole 2002). *Scolopendra gigantea* is famed for

Scolopendra gigantea and relatives are not terribly picky. This female grabbed a big piece of sausage and carried it up the screen to eat it. © David T. Anderson

reports of specimens catching and eating bats in caves (Molinari et al. 2005). *Scolopendra* have been encountered feeding on small snakes in nature, such as a two foot (50 cm) Sonoran gophersnake *Pituophis catenifer afinis* found being attacked and chewed on by a 7″ (18 cm) *Scolopendra heros arizonensis* (Babb and Busboom 2013), but they are more commonly found in snake stomachs (snakes can swallow centipedes larger than themselves). Centipedes found in bird's nests tend to be tiny and specialize in eating mites and parasites. Most centipedes readily scavenge dead animals. In captivity, when hungry, they will feed on processed meats such as hamburger and hotdog, but also banana and other fruit. Some of the

Scutigerid eating an adult mantis (*Statilia pallida*) in the rainforest, island of Panay, Philipines. © Christian Schwarz

A relatively large, soft-bodied scorpion can be eaten. *Scolopendra heros heros*. © David T. Anderson

Geophilomorpha feed on decayed leaves and wood or, rarely, live plants (Nichols et al. 1971). The house centipede *Scutigera coleoptrata* is commonly encountered around crumbs under the toaster oven or microwave. However, it will not eat bread or dog food in captivity and is waiting for prey attracted to the crumbs.

Even the most enormous centipedes have a difficult time with large prey. Most four-inch mantids easily dispatch a four-inch cockroach, while a four-inch centipede has trouble with food larger than an inch (earthworms and dead food are exceptions). In general, it is best to introduce prey not much more than twice the size of the centipede's head (or much greater than the width of its body). Of course, some hobbyists become interested in centipedes because of their famed ability to eat mice. It is possible to feed pinkie mice to most Scolopendromorphs over 5″ (wait a few weeks or more to let the centipede grow hungry). However, some countries prohibit feeding live vertebrates to invertebrates because of animal rights issues. Such food should not be encouraged and is not necessary live. Frogs, lizards, and far too large of cockroaches can also be consumed. Centipedes can, at times, take out surprisingly oversized prey, but regularly feeding by exception is not a good method because it decreases chances of survival, significantly for early instars. It may be a poor analogy, but one would not starve a cat or dog for long periods to see how large of prey they might be capable of taking out. Also, refuse removal is important. Oversized, uneaten prey remains should be removed within hours. Centipedes can also be put at risk of damage when offered overly large prey.

Heavily-armored scorpions like this *Heterometrus* species can be taken when teneral. *Scolopendra gigantea*. © David T. Anderson

Scolopendra gigantea eating a small, soft-bodied scorpion. © David T. Anderson

A teneral lobster cockroach, *Nauphoeta cinerea*, being consumed by *Scolopendra dehaani*.

Centipedes are perturbed by thick armor. The maxillipeds and second maxillary palps cannot chew, while the first maxillary palps covering the mandibles are not terribly powerful and terminate in what looks like small molars. These "molars" are tiny compared to the overall size of the centipede and are simply incapable of pushing something large and hardened through to the mandibles. If a thick-shelled beetle, scorpion, or cockroach is offered the centipede ignores the item or drops the prey after attempting a bite or killing it. Commonly cultured feeder beetles like adult pea weevils, dermestids, or mealworm beetles are generally worthless as feeders for centipedes. Buthid scorpions often have large soft areas of the exoskeleton and can be eaten, but centipedes have no chance against hard-shelled species. Most cockroaches do not have exoskeletons as thick as the popular hissing cockroaches. Still, small nymphs of the most heavily armored species have very thin, soft shells. Soft or hard-shelled cockroaches of any size are extremely palatable to centipedes if offered while teneral. The soft, white treat is beloved and makes a great meal for a sickly or favorite centipede. Decent size cockroach colonies have teneral individuals present regularly.

Caging affects the prey type to be chosen. Burrowing prey can be a problem, especially in a large cage, or one with deep substrate. Unless the keeper has time to make sure each prey item is killed and consumed, some of these may come back to bite one in the terminal segment. Climbing prey can likewise be a problem since centipedes cannot climb smooth surfaces, and almost never swing their bodies around to catch creatures above the surface (in contrast to the bat video). In larger or decorated cages, prey must be chosen carefully since issues are not always obvious.

Older juvenile and adult centipedes are rarely sensitive to the quantity and frequency of feeding. One cricket, once a week, for a four-inch centipede is plenty. Overfeeding causes more trouble than underfeeding. If the centipede is not hungry it will usually kill energetic prey like crickets without eating them. Unless the prey has plenty of space and hiding places, cut down on feeding or remove prey that has not been consumed within a day or two. Decomposing prey can poison the centipede and excessive live prey noticeably stresses the animal. Evidence of excessive rotting prey is easily hidden by the centipede's burrowing activities which bury but do not solve the problem. Adult females preparing for a brood should be fed as much as they will eat until they lose their appetite but then should not be fed again until the offspring have dispersed. Young can be fed every day to speed up growth, but most stay healthy when fed once a week. Sometimes a new import will arrive and seem to eat everything thrown at it, even when fed twice a day. Every last bit of the prey appears to be consumed. Feeding such an animal until it can eat no more (instead of weekly) will result in problems unless the cage substrate is changed out and discarded weekly. Overfed animals produce a lot of undigested waste and it can be very difficult to identify and remove all frass particles unless the substrate material is white.

A variety of foods are available to centipedes in the wild. Variety in captivity is not just for variety's sake—two or three cultured prey items should be plenty. A varied diet improves health, growth, and reproduction, since any single prey item lacks some nutrients. Foods such as fruit flies, fruit, and isopods lack some basic nutrients and cannot be used alone, but are useful for certain stages, clean-up, or to induce feeding. Cultured items go through boom and bust cycles so a little variety also helps to ensure there is surplus available for feeding.

These *Scolopendra hardwickei* are approximately the same age. The smaller animal was maintained by a different keeper and not fed very often. © David T. Anderson

Worms—Various worms, primarily Oligochaeta, from common earth worms to African red worms and nightcrawlers fall under this food category. There are also commonly cultured "white worms" which can be used to feed early instar scolopendromorphs. Nematodes can be reared for small stone centipedes.

Pros: Earth worms are found everywhere; all that is needed is a shovel and some time. Worms can be purchased at most convenient stores and bait shops. Even bright green or orange nightcrawlers are available. Worms are high in protein and likely a part of the natural centipede diet. Uneaten worms do not harm centipedes. Centipedes can eat worms larger than themselves.

Cons: The major problem with worms is it can be difficult to tell if the worm was eaten, escaped, died, or has pasted itself with substrate. Worms tend to go straight down into the substrate and may not be eaten right away; though it is neat to watch a centipede carefully tugging worms out of the dirt. Worms are able to climb glass and escape through tiny cracks. Most experience high temperature sensitivity. Many earth worms "melt" when the temperature gets much above 80° F (>27° C). If small worms are needed for feeding baby centipedes, they need to be cultured.

Rearing: A plastic shoebox makes a good, inexpensive culture container. General rearing guidelines for earth worms has not change much since Butler (1856) described rearing red worms as feeders for tropical fish. The box can be filled with moist compost or potting soil. Dog biscuits, fish flakes, leaf mulch, and table scraps can be tossed on the substrate for food. Most can be reared in a cool garage but freezing will destroy a culture. During hot periods, the container should be stored in a refrigerator or on the basement floor. However, African red worm cultures (commonly available for feeding tropical fish) are tolerant of higher temperatures (and probably intolerant of extreme cold).

Crickets—The house cricket, *Acheta domesticus*, is the most common feeder. A few other species have shown up at pet shops in recent years due to a species-specific virus that decimated cricket farms a few years back.

Pros: Feeder crickets can be purchased at nearly any pet shop and through the mail. They

Scolopendra hardwickei eating a cricket. © David T. Anderson

food. Water should be provided to the growing crickets by keeping an area of the substrate wet. It is necessary to use a screen or otherwise ventilated cage or the crickets will die within a few hours to a few days. Additionally, a 60-watt incandescent light bulb located directly atop the cage is a necessity unless the room temperature is above 75° F (24° C). Without this source of heat it is often impossible to hatch out a single cricket, but it is also important to keep the egg-laying media from drying out. Growing and reproductive crickets will use the light for basking.

Scolopendra gigantea eating a cricket. © David T. Anderson

can be purchased in numerous sizes to accommodate the size of the centipedes being fed. Crickets are active and quickly draw the attention of hungry centipedes.

Cons: Uneaten crickets are dangerous to molting centipedes. Crickets can be relatively expensive over the long run and culturing requires far more space and effort than other feeders listed here. Crickets are excellent jumpers and escape often into the home where they can chew on books, food, and other personal items. The sound of a loud male chirping away every night interrupts the sleep of many hobbyists. Small crickets, especially pinheads, are very difficult to handle and those that do not jump away are often squashed by the act of picking them up. (Pinheads can be placed in a refrigerator for ten minutes to make handling easier.) Dead crickets are prime for infestation with phorid flies and grain mites.

Rearing: Adult female house crickets lay a few hundred eggs in any potting soil or dirt that is an inch or more deep. The dirt should be kept moist until tiny pinhead crickets hatch out after a few weeks. The nymphs and adults do well on many foods such as most leftovers, old fruits, many vegetables, and dry dog or cat

COCKROACHES—*Nauphoeta cinerea, Blaberus discoidalis* "discoids," *Blaberus craniifer* "death's heads," *Blaptica dubia*, *Blatta lateralis* "red runners," *Schultesia lampyridiformis* "firefly mimics," *Dorylaea orini*, etc. Cockroaches as a whole come second only to crickets as the most common and proven captive centipede food. Members of certain burrowing genera, specifically *Gyna* and *Pycnoscelus* species, should never be used.

Pros: Rearing feeder roaches is inexpensive, takes little time, and most people find it rather easy. Cockroaches live inside the cages until the centipede is ready to eat them. There are numerous types and sizes of cockroaches

Nauphoeta cinerea offer ease of rearing, an excellent size range, soft bodies, and limited burrowing.

Scolopendra heros heros eating a somewhat teneral *Blaptica dubia* nymph. © David T. Anderson

An immature *heros* x *castaneiceps* hybrid trying to eat an adult male *Blaptica dubia*. © David T. Anderson

A delicious *Dorylaea orini*.

available for feeding centipedes. *Blaberus*, *Blaptica*, and some other genera are unable to climb glass. This makes them much easy to contain and they stay near the bottom where the centipede can catch them. Other useful feeder cockroaches such as *S. lampyridiformis, N. cinerea,* and *D. orini* can climb nearly any surface and may escape feeding, but they are thinner shelled and do not burrow. Small cockroach species, when born or hatched, are a good size to feed most young centipedes and are easier to handle than other small prey. Most cockroaches other than *Gyna* and *Pycnoscelus* do not bother molting centipedes. Some of these feeders are pretty in their own right and none chirp.

Cons: Cockroaches tend to be more expensive than crickets and are sold mostly as starter cultures. Culture starts of the more interesting species can cost more than the centipede. Large cages are often necessary to rear large numbers. Glass-climbing species may hang out at the top of the cage where the centipede will never reach them. Cockroaches hide and may avoid getting eaten for months.

Teneral firefly mimic cockroach gone in fifteen seconds.

Rearing: Cockroaches are often kept in five-gallon buckets, ten-gallon aquariums, or fifty-gallon trashcans. Egg cartons or large pieces of bark are placed in the cage for climbing, molting, and hiding. A water dish containing a sponge of the same or greater diameter and height to prevent drowning should be provided (small sponge pieces, gravel, or sticks placed in the water may prevent drowning but lead to rotting accumulations of dead bodies, body parts, and old food infested with maggots and fungus). Nearly any human or pet food can be used to feed roaches. Fresh fruit should be offered from time to time to the live-bearing species (i.e. *B. discoidalis* and *N. cinerea*) or they may produce fewer or no young. Egg-laying species (i.e. *D. orini* and *Blatta lateralis*) should be misted from time to time to keep the eggcases from drying out. Cages that contain glass-climbing species should have the upper two inches coated with petroleum jelly to help prevent escapes when the lid is removed.

Small firefly mimic cockroach being consumed.

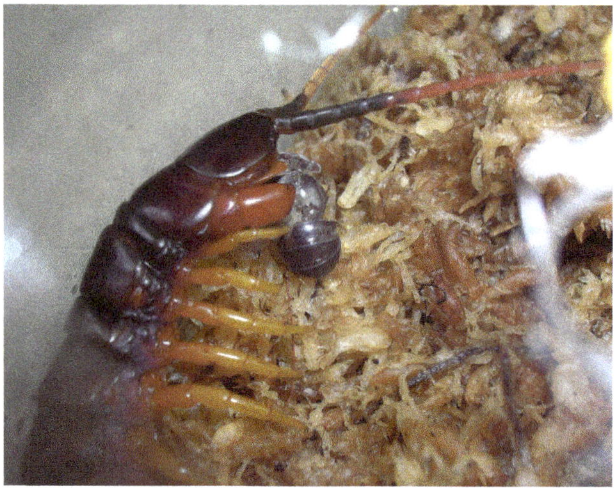

Armadillidium vulgare being eating by an immature, captive-hatched *Scolopendra gigantea*. © David T. Anderson

Isopods—Terrestrial isopods (those that roll into a sphere are known as pillbugs) are found worldwide. There are at least 112 species in the U.S., of which 36 are adventive (Jass and Klausmeier 2001). The most commonly encountered include *Porcellio laevis*, *Porcellio scaber*, and *Armadillidium vulgare*, all adventives from Europe. Centipedes more rapidly eat species that are able to roll into a ball, despite the defensive mechanism. In recent years, a few extremely tiny species have become readily available for keeping vivariums clean and have been used as a partial diet for feeding small *Scolopendra*.

Pros: The hobbyist can collect these small invertebrates under logs and leaf litter anywhere. Isopods are among the most common terrestrial invertebrates and are even found in the desert, or in the dead of winter (under a

There are some colorful isopods commonly available as culture starts, including the zebra pillbug *Armadillidum maculatum* (left) and orange form of *Armadillidium nasatum* (right).

log beneath two feet of snow). They are part of most centipedes' natural diet. Uneaten isopods are very unlikely to harm a molting centipede. They eat leftovers from the centipede's meal, which helps to reduce phorid and mite problems. Like centipedes, isopods are intolerant of dryness and dried-up isopods can indicate poor conditions before the centipede desiccates. Centipedes do not seem threatened by isopods and tend to only kill them when hungry. Young centipedes can eat relatively large isopods.

Cons: Many centipedes will not eat isopods unless quite hungry. They are often used more as "clean-up crews" or back up prey when feeding has been forgotten. Culture starts can be somewhat expensive, though mostly for colorful species or cultivars. Unlike other terrestrial animals, isopods excrete ammonia as gas rather than bound up as uric acid or crystals. Though highly unlikely, excessive numbers in a poorly vented cage might poison themselves and the centipede more quickly than other feeder animals, but there should never be hundreds of feeder animals in with a single centipede. Terrestrial isopods seldom exceed 1/2" (0.5 cm) and many would be required to feed a really large centipede. Small cultures cannot produce great quantities because the turnaround time is many months.

Rearing: The lid should be well-vented for pillbugs, though standard isopods require less ventilation. The culture container can be filled with 1" to 2" (2.5-5 cm) of moist compost, leaf mulch, or soil. Place a layer of dead, brown hardwood leaves on the surface. Isopods feed on bulk animal foods and table scraps, but decaying leaves are an important part of the diet for a thriving culture. Sexual maturity is reached before they are half-grown. Adults continue to molt and take a few years to reach maximum size.

Yellow Mealworms—*Tenebrio molitor*. Mealworms may be the oldest feeder insect. At least one aquarium book from the 1920s discusses rearing mealworms as food for tropical fish.

Pros: Mealworms are among the most accessible foods and can often be found at grocery, convenient, fishing/tackle/bait stores as well as pet shops. Hatchling larvae are half the mass of a fruit fly and can be fed to the tiniest baby centipede. They are inexpensive and easy to rear. Relatively huge cultures can be reared in minimal space.

Cons: A large number of full-grown larvae are needed to provide the nutrients for even a medium-sized centipede to make it to the next molt. Mealworms can be dangerous to molting centipedes. They immediately burrow into the substrate and can stay unnoticed for a long time, but very seldom take revenge on a molting centipede. Mundoworms or super mealworms, *Zophobas morio,* are much larger and more vicious. They may even bite a centipede that is not molting and are not recommended.

Mealworm culture, *Tenebrio molitor.*

Scolopendra dehaani chewing on a banana slice.

Rearing: Any glass or plastic container with lid will suffice as a cage. Fill the container a few inches from the top with plain, uncooked oatmeal or dry dog food. To provide moisture, without causing the oatmeal to mold, small pieces of potato can be placed on top of the oatmeal and replaced every week or two. Misting the top of the oatmeal every few days is another way to provide moisture. Even if it is not damp enough to mold, moisture can lead to massive grain mite infestations so cultures are often kept on anti mite paper to prevent mites from getting into clean cultures (commonly used with fruit flies cultures as well). A solid lid can be used on the rearing container but then misting would cause the substrate to mold.

Pinkie Mice—There is no hair yet on newborn mice to cover the pink skin, which gives them their name.

Pros: Available at many pet stores. Feeding impresses how large centipedes are; not many invertebrates are massive enough to eat vertebrates. Pinkies are feeble and may not harm the centipede but only frozen/thawed are suggested.

Cons: Uneaten carcass portions should be removed immediately to cut down on unpleasant smell and byproducts of decomposition. Pinkie mice are not likely a part of the giant centipede's natural diet. When it comes to larger or living mice, consider the National Geographic video clip of a grasshopper mouse tearing leg after leg from an adult *Scolopendra heros arizonsis* until the massive centipede dies.

Bananas/Fruit—Though an odd food for a predatory animal, banana may provide some nutrition not available in other foods. Other

Scolopendra heros arizonensis feeding on a chunk of fruit jelly. Fruit jelly is easier to chew than fruit and can be useful in nursing a sickly animal.

fruits including cantaloupe and apple have been used, as well as slices of insect jellies.

Pros: Harmless, extremely cheap, and easy to find. Fruit may be used for a short time when regular food is not available. It can be used to nurse back an emaciated or sickly centipede that is too exhausted to catch or chew up prey.

Cons: Except for Scutigeromorphs, all centipedes will nibble on banana when hungry, but only a few species really gobble it up. If used as the only food, most centipedes would stop growing and die prematurely. Bananas or other fruits are not likely a part of the natural diet of large predatory centipedes.

FRUIT FLIES—*Drosophila* species are available in flightless mutations.

Pros: Harmless, food for small and young centipedes that evokes excellent feeding response.

Cons: Cannot be used as the only food or immature centipedes will not grow. The largest fruit flies are small. Keeping continuous cultures takes work. They do not live long in centipede cages.

Rearing: A 32 oz deli cup with coffee filter lid is a good culture container. Medium consists of 1/2 cup instant potato flakes and 1/2 cup water—varying somewhat by ambient humidity. Yeast, vinegar or an apple slice can be added to promote egg laying. Keep temperature above 75° F (24° C) for best productivity.

Fruit fly culture, *Drosophila melanogaster*.

"GIANT SPRINGTAILS"—There are a number of springtails commonly cultured for use in terraria but only the common species sold under this trade name is really useful for centipedes because the maximum size is relatively large (as big as a very small *Drosophila*).

Pros: Springtails are very tiny and are a proven starter food for Scutigeromorpha, various small Scolopendromorpha, and have been eaten by Lithobiomorpha and Geophilomorpha. They are useful as clean-up crews. Cultures require little upkeep and can stay in the same container for a decade.

"Giant" springtail culture.

Cons: A few months to half a year are required to establish a large culture from a starter. Since they are very tiny, they are only used for small and early instar centipedes.

Rearing Instructions: Springtails feed on a variety of different foods but are normally raised on fish flakes, dog food, and fruit. Very little food is required. Springtails are kept on a shallow layer of potting soil, compost, or charcoal and do best if kept in medium to high moisture. They require little ventilation so a few pinholes or the normal cracks around a plastic shoebox lid provide plenty of air. Remove any small brown mites (*Hypoaspis* sp.) that appear in the culture container or the production can be minimal or negative. (Predatory mites prefer to eat other mites rather than springtails, but they may eventually wipe out a springtail culture under some circumstances.)

ORDER SCOLOPENDROMORPHA

INDICATIONS: Primarily with 21 or 23 leg pairs and segments; the Brazilian species, *Scolopendropsis duplicata* Chagas, Edgecombe, and Minelli, 2008, has 39 or 43 leg pairs and segments.

Scolopendra is a Latin word, meaning *centipede*. In this order the enthusiast is concerned almost entirely with the Scolopendridae, with its two subfamilies the Scolopendrinae and Otostigminae. These are often large species and have four eyes on each side of the head. *Scolopendropsis duplicata* can exceed 3″ (>7.5 cm) but it is very thin like a geophilomorph (with thirty-nine or forty-three leg pairs and segments, by gender). Other species such as *Scolopendropsis bahiensis* (Brandt, 1841) have twenty-one or twenty-three body segments and more usual proportions. The two species of Old World Mimopidae have a single pair of eyes, while the remaining families (Cryptopidae, Plutoniumidae, and Scolocryptopidae) have no eyes. The family Cryptopidae contains some interesting species, but representatives are generally small and reddish yellow in color, so a quick count of the legs is needed to make sure they are not stone centipedes.

SEXING AND MATING

This order of centipedes is often between difficult and impossible to sex externally. The females may be heavier and fatter but there is no notable external difference. Because of the difficulty of gender determination, notes on egg laying, spermatophore production and mating attempts should be kept with each animal. Sexual dimorphism is uncommon, although it is seen in some species. The terminal legs are used during mating and can be shaped slightly differently between males and females of a few species, including *Scolopendra morsitans*, *Parotostigmus rex*, and *Otostigmus scaber*. Differences may also be seen in color, such as the much darker blue and orange of the male Tanzanian *Scolopendra* sp. "Neon Blue." In *Scolopendra hainanum* the females have banded legs while the males' legs are usually solid yellow.

For most species we need to find a way to check the gender which is not external. It is possible to safely anesthetize larger specimens in order to carefully push out the genitals for sexing. The male has an organ used in spinning the web structure where the spermatophore will be laid. It sounds very easy but the spinneret is retracted and deflated when not in use. Identifying male and female internal genitalia can result in absolute determination of gender, but there is certainly some variability between specimens and species. If a number of adult specimens are available or one is familiar with the genitals of a specific species, this is the only method capable of providing a way to identify gender with certainty.

Everting genitalia, anesthetized specimen, *Scolopendra heros castaneiceps* male. © John Skierski

Scolopendra heros castaneiceps, male genitalia. © John Skierski

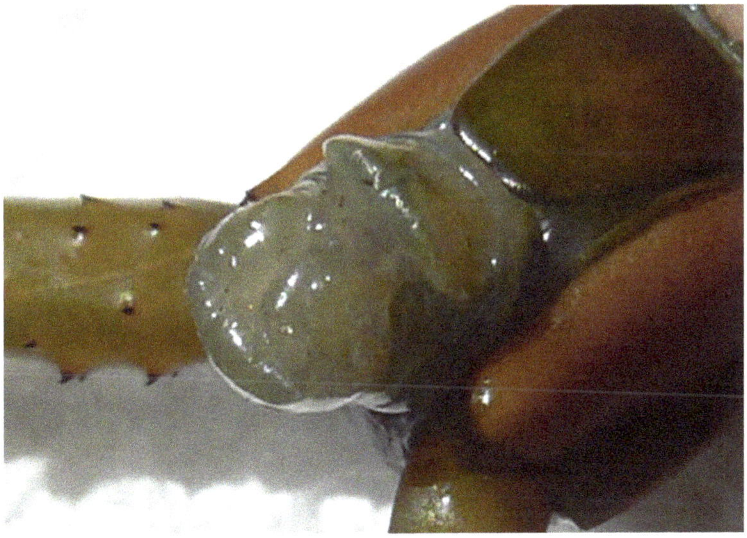

Scolopendra heros castaneiceps, female genitalia. © John Skierski

Scolopendra dehaani, Female genitalia. © John Skierski

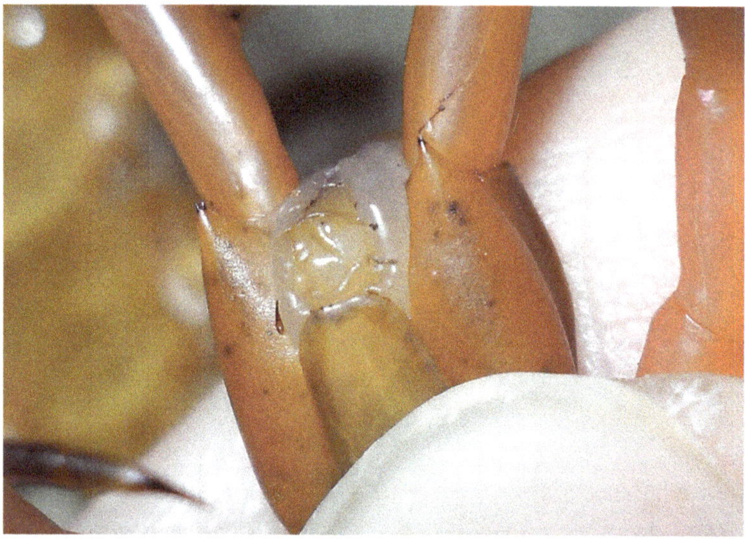

Scolopendra dehaani genitalia, gender—? I squeezed this thing like a tube of toothepaste. It is not always easy.

Male on left, female right. In most species it is not easy to differentiate male and female externally.

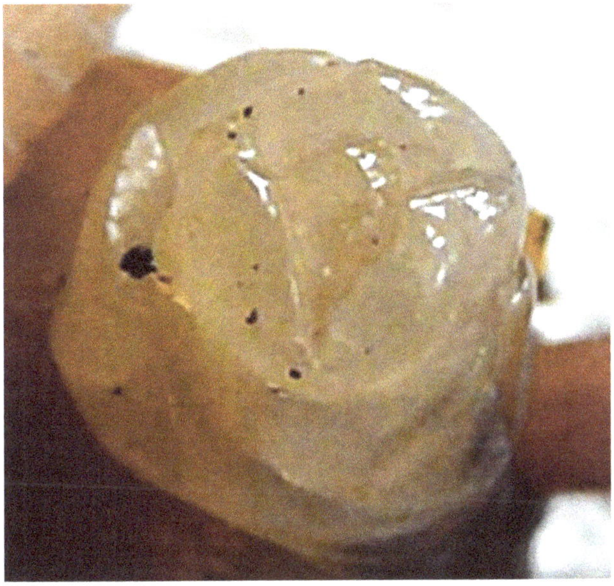

Scolopendra subspinipes mutilans
Red Leg, female genitalia. © John Skierski

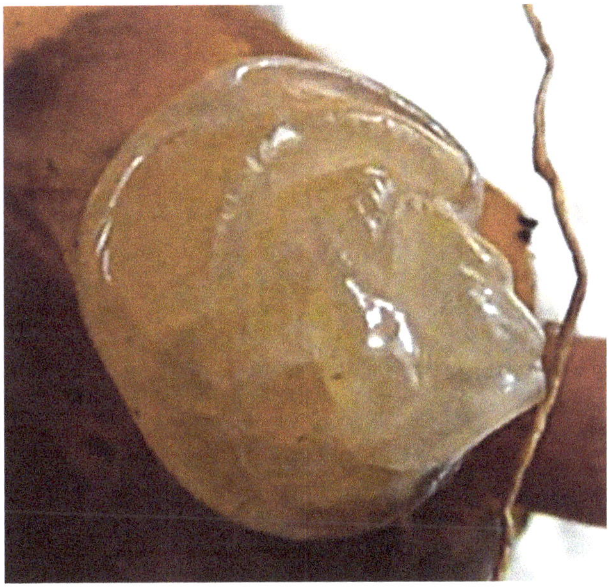

Scolopendra subspinipes mutilans
Yellow Leg, female genitalia. © John Skierski

ANESTHETIZATION PROCEDURE FOR SCOLOPENDROMORPHA GENDER IDENTIFICATION

When attempting to sex animals from this order, the most common method is to utilize carbon dioxide (CO_2) gas within a small chamber where the centipede will be contained. There are many types of equipment and other possible gases or fluids available to the keeper. Short-term, near drowning does not result in the immediate death of healthy animals, but eversion of the genitalia is more difficult while eventual lethargy and death can result, especially in the case of less healthy specimens. Here we will stick with inexpensive and readily accessible items that do not require permits and should reduce chances of accident through miscalculation. The following equipment and methods have been used successfully to bring approximately thirty animals of various species (*Scolopendra heros*, *Scolopendra dehaani*, *Scolopendra subspinipes*, *Scolopendra polymorpha*, *Alipes*, and *Ethmostigmus*) in and out of unconsciousness safely (Skierski pers. comm.).

Materials

· One soda stream or whip cream dispenser that uses CO_2 cartridges.

· CO_2 cartridges/chargers, standard 8 gram. Note: cartridges of this size and used in conjunction with even a small container are only capable of knocking out a centipede for a couple of minutes. It is necessary to have all equipment in place, camera focused, etc. to allow maximum time to photograph genitalia. Alternately, if you have access to larger cartridges or dispensers or a better means of measuring how much gas is being used it will allow for more time but can involve more risk.

· Aquarium airline tubing (must fit snugly over dispensing tip).

· 32 to 40 oz container, whatever minimum volume capacity can be used safely. Tall deli-cups or plastic jars may be used with smaller centipedes. Ziploc or Gladware style rectangular containers that measure roughly 5″ L x 4″ W x 3″ D (13 cm x 10 cm x 7.5 cm). Something

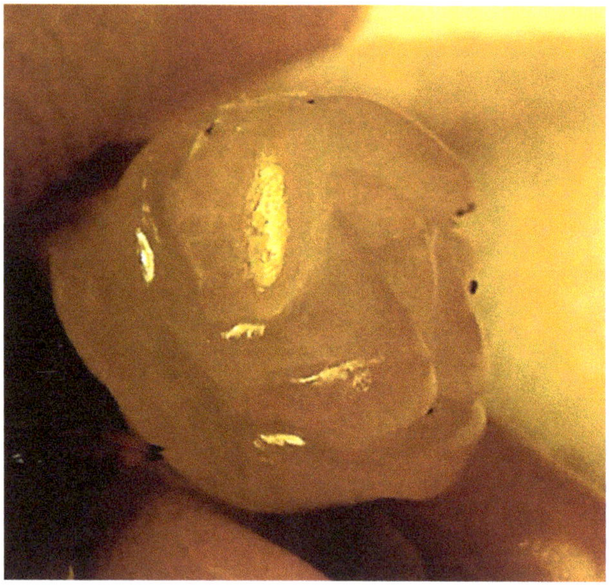

Scolopendra subspinipes, Malaysia, female genitalia. © John Skierski

Scolopendra subspinipes , Barbados, male genitalia. © John Skierski

more oblong they could naturally crawl into, such as an oversized cigar tube would be easier to work with (no transfer and less volume for gas) but try getting a centipede to enter a hole that narrow.

· Duct tape or rubber washers that will allow for securing the tubing to the container.

· A camera capable of taking clear photos of the genitalia. A macro lens with additional lighting is highly recommended, although many point and shoot cameras now have reasonable capabilities to capture the necessary image zoomed in or held close (one of the *S. dehaani* shots is an inexpensive Nikon L22 Coolpix set to macro with low exposure compensation to adjust for the closeness of the flash). If the centipede is much under 4″ (<10 cm) body length the magnification of the camera used here is not high enough for conclusive pictures (Panasonic Lumix FZ50 with Raynox DCR-250 macro attachment).

Assemble the "gas chamber" apparatus by attaching one end of the tubing to the dispenser tip. Cut tubing to the desired length—a few feet (~0.6 m) is plenty. Cut one hole in the container, about the diameter of the tubing and attach the other end here. Using copious amounts of duct tape or a rubber washer, secure the tubing entering the container so that there are no gas leaks and so it does not blow loose during use. Make a small hole (or a

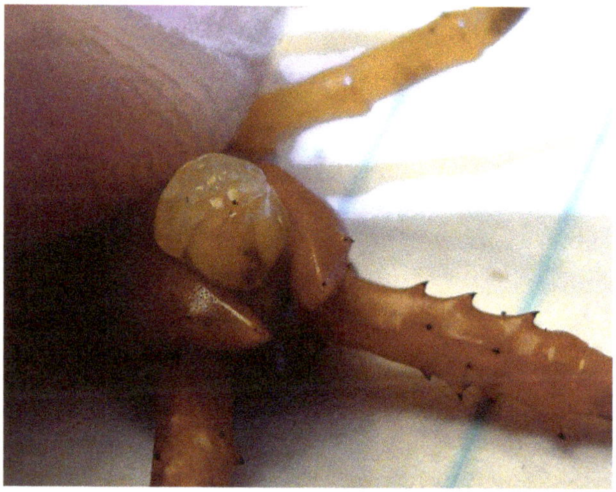

Ethmostigmus sp., Tanzanian "Giant Yellowleg," female genitalia. © John Skierski

series of thumbtack holes) in the lid of the container to allow for air escape as the gas fills the container (CO_2 is heavier than the air it is replacing).

Transferring the Centipede

Getting a healthy giant centipede into the chamber without being bitten is no easy feat. Keepers who are overly nervous, not nervous at all, or unfamiliar with centipede behavior will need to devise a better plan to safely move the centipede or give up on this method of gender determination. One option is to transfer the animal from its enclosure to the container using a large deli cup. The smooth sides of the container will allow the animal to slide into the container without damage and should allow enough time to hold the lid over the top and snap it in place. Be sure to avoid catching body parts in the lid. The container can be rocked back and forth to keep the animal at the bottom as the lid is secured. Centipedes grab onto and walk up edges so this part can take some patience and a steady hand (holding forceps).

Eversion and Photographs

With centipede in place it is time to pump CO_2 into the chamber. Avoid unleashing all the gas at once or the lid can be blown right off the container holding a large, pissed-off centipede. After the cartridge is empty you can tell when the animal loses consciousness as it will stop moving completely and remain limp. Within 30-45 seconds after the cartridge is empty, if not sooner, the animal will stop twitching or moving completely. Once the animal is appreciably limp, it is time to take it out.

With great care remove the animal quickly and line up the last segment at the terminals dead center in front of the camera, ventral side up. This is where the genitals will be everted. Gently roll a finger or spoon across the second to last sternite, towards the terminals, until the genitals are fully everted. Little pressure is needed to do this with a fully unconscious centipede. If there is resistance, additional CO_2 may be needed. The genitals will appear to inflate like a small balloon. Take as many photographs as you can while the animal remains subdued. An angled profile will allow visibility of the entire anatomical detail—shots from above are usually inconclusive. Make sure to write down which photos are for which specimen.

I have never had a centipede not wake up within five minutes, even when using two cartridges. Nor have I had one die soon after or show any negative effects. I have had them wake sooner than expected and grip my hand—very exciting with a large specimen! But, that is not the norm. They start out very groggy but are fully functional within fifteen minutes. Often they will attack prey or drink within minutes of recovery.

So now that we have some clear photographs, what are we looking for without getting too indepth into genital anatomy? Fortunately, most large male adult centipedes, and especially from the genus *Scolopendra*, will have obvious appendages that resemble snail's eyes sticking out from the upper part of the everted genitals just below the genital sternite (an obvious layer covering the central area of the genitalia when everted). Also, only males will have a second genital sternite partially covering the first. Female's genital sternites are often markedly rounded above the main genital area.

Keep in mind that going in and out of consciousness has visible signs. As the centipede enters unconsciousness the legs often twitch rapidly at the ends. As it begins to regain consciousness the claws begin to hold onto things while the legs stay limp. Shortly after, the maxillipeds will be opened and closed a little and the legs flex slightly. The animal will not simply begin to act normal right away, but it will begin to grab on.

If the genders are not easily determined externally and genitalia extrusion is not possible or inconclusive, a good estimate on the gender can be made using relative size and bulk. Females tend to be bulkier, not necessarily longer. This can work well for some species, but will prove difficult or impossible for others. The Tanzanian "Neon Blue" is quite easy to sex by bulk, whereas *Alipes grandidieri* always look thin even when about to lay eggs. *Ethmostigmus trigonopodus* can appear morbidly obese whether male or female. *Rhysida longipes* and some of the *Scolopendra*, especially *S. alternans* and *S. heros* can be easy to sex in this way, but it is only possible to pick out the largest and gravid females with any consistency. If one has ten adults it is easy to pick out a few big females with accuracy but not conclusively sex the rest. Bulk is not a perfect method, but a good last ditch effort before putting specimens together or for picking out a specimen that is likely to lay eggs.

The very last sexing method is post-introduction. The idea is to observe and record the reactions of adult specimens after they have already been placed together. Of course it would be necessary to have adults of both sexes and only positive results could be conclusive. Males will usually flash the terminals at the other animal in display and, if the other is a female, she will tap his terminals with her antennae. If two males or two females are placed together they often face terminals towards each other before a full-on fight (Anderson 2008).

Choosing and conditioning adults for mating can take some time even after specimens have been conclusively sexed. Since there are no signs of maturity, animals should be full-

Female *Scolopendra hainanum* with banded legs. © D. T. Anderson

Male *Scolopendra hainanum* with unbanded legs. © D. T. Anderson

size and should not have had a recent molt accompanied by a notable size increase (with very rare exception, sexually mature animals grow minimally, if at all, in further molts). Even if they are the correct gender and fully mature, hungry specimens can choose to eat rather than mate. The pair should be well-fed and females, especially, should not look thin. Mating should not be attempted with animals that have no appetite, as they may be sick or preparing to molt.

The only proven method for mating centipedes is placing specimens together. In the rare

Scolopendra viridis (and lost egg).

Scolopendra viridicornis.

Scolopendromorph, Western Australia.
(CC) Greg Schechter

case of Scolopendromorpha like *Scolopendra subspinipes mutilans,* the adults and large immatures can be kept together with relative safety in tiny, nearly empty cages. The Tanzanian *Ethmostigmus* imports can be kept together safely for years with available shelter. Others such as *Alipes* may stay together for months but do cannibalize from time to time. For most species, however, receptive pairs can only be kept together for a few hours to a day or one will be eaten. If they are not receptive or the wrong sex, one can be killed within moments (the death of both is highly unlikely).

Hands-on, short-term mating cages can be relatively small and still successful. © David T. Anderson

Breeding procession, male in front. *Scolopendra heros heros*. © David T. Anderson

The introductory enclosure should be large enough to offer avenues for escape or courtship. A large, smooth-sided container with a few inches of substrate and some hiding spots can be used to introduce pairs. Some shallow cork bark or rock leans are useful for providing the opportunity to form spermatophore webs. Forty- to fifty-gallon tubs with lids are inexpensive and can be found at any department or hardware store. The height of the tub should be greater than the total length of the longest centipede (including rear legs and substrate depth). At first, long forceps can be at the ready in order to separate them at the first sign of aggression. A spray bottle set on stream can work more quickly, but forceps will still be needed to move specimens. Hopefully they will begin tapping with their antennae and cautiously circle each other. Pairs can be left together for a few days if they immediately run and hide, but keep in mind, centipedes are cannibalistic, and not necessarily after a successful mating. Mealworms and isopods should be put in the cage as food. Avoid large crickets because they can disturb and irritate the centipedes during this crucial period.

Mating often occurs within the first hour after initial contact. Glass terrariums with objects leaning against the glass can often give the viewer the benefit of seeing what is going on. The male initially exposes his terminals towards the female and if she is receptive she taps on them with her antennae. He shakes the terminals in response and she taps again. This happens every few seconds for many minutes to as long as an hour. He will then walk in search of a good spot while keeping in contact with her through terminal taps. If he loses her,

(Top to Bottom) Mating procession from above, *Scolopendra subspinipes*.
© John Skierski

he will find her and shake his terminals at her until she responds. If she is not going the right direction he will back up and grab onto her body with the terminals to correct her course (remember they are both facing forward). This may resemble an attack but with the male's rear legs. Once a good spot is chosen he begins to spin a web tunnel where the spermatophore will be placed while he keeps in contact with her using his terminals. This takes at least a few minutes but it is not a terribly complex web. The structure is usually formed below a shallow, flat surface or overhang. Once the spinning is done he deposits a rather large, white spermatophore and then walks forward. The female follows him through the tunnel and picks up the spermatophore with her genitalia, located in the last body segment. When lucky enough to witness a successful mating, it is best to separate the centipedes as soon as the spermatophore has been picked up to avoid them harming each other.

In the vast majority of species it is very important to separate the adults at the slightest sign of aggression during the mating procession, or one will most certainly be eaten. Tapping and moving the female with the terminals are not aggressive, while attacking with the head or biting are bad signs. However, in the case of *Scolopendra hardwickei*, it seems the male may bite the female as a normal part of early courtship, but she does not react (Anderson and McMonigle 2013).

Though mating can be challenging and dangerous to livestock, one success can really pay off. Captive-mated centipedes are able to produce three consecutive, fertile broods from a single mating (McMonigle 2007), however, sperm storage of eighteen months or greater is highly unlikely and has never been documented. One female *Alipes grandidieri* which produced three consecutive broods did not molt between or before them. *Scolopendra heros* and *S. polymorpha* commonly produce viable young the following reproductive season after capture or mating, sometimes twelve months or more and following a molt. Still, any eggs produced the second season are not fertile and will not develop without introduction of a male. Sperm retention is not affected by molting. This is very different from tarantulas and whipspiders which shed the sperm storage compartments after a molt, but mature female millipeds can also molt and retain sperm. In addition to *S. heros* and *S. polymorpha*, *S. alternans* and *S. viridis* have also molted and produced viable young afterwards without being reintroduced to a male (Skierski pers. comm.).

This female waits while the male spins the web with his large spinning organ.

The male scolopendromorph produces a large spermatophore and deposits it on silk threads. Spermatophores look like white slugs and are generally only deposited near the end of a mating procession. Before formation, the male takes on the task of persuading the female to take up the spermatophore by tapping her and getting her to follow him to a good spot for the spermatophore. It had been suggested that the male might leave spermatophores in places for females to find. However, the associated mating procession is important for her

The male deposits his spermatophore on the web structure. The spermatophore is large and looks like a white slug. © David T. Anderson

Success, the female accepts his spermatophore. © David T. Anderson

Female *Scolopendra heros heros* checking out the male.

interest. So far, no one has witnessed (let alone photographed) a female accepting an abandoned spermatophore. Gamete broadcasting is known in corals and conifers but they employ ocean currents or wind because they cannot move. Broadcasting of the male gametes would require consistent population density over millions of years, precise timing and high gamete production. Centipede spermatophores are seldom encountered in captivity and then only immediately following courtship.

There are two "alternative" centipede mating ideas. At one time these were considered real theories.

1. Barrier method: a cage is designed with four or more compartments and a centipede placed in each. The lower portions of the dividing walls (usually drilled plexiglass or eggcrate) are riddled with holes a bit narrower than the trunk of the centipedes. Due to indeterminate gender, multiple compartments are used to increase the number of shared walls. Only two compartments are needed with specimens of known gender. The theory is the antennae and legs can touch for mating purposes, while the centipedes cannot get through to eat each other. The cage must be carefully designed to prevent the animals from jumping compartments or escaping. On the face, it sounds reasonable, but is not demonstrated and the required mating procession is likely impossible.

2. Rotation method: centipedes are kept in separate cages and rotated between cages every month or two. The gender is usually unknown, so as many animals as possible are used to increase the chances of a female introduced into a male's cage. The idea is the male will leave a spermatophore somewhere in the cage and when a female runs across it she will take it up into her genital opening. This method takes very little work but is a myth. There has never been a single demonstrated case (male spermatophores are large and visible even for

On substrate too shallow or too loose to form a brood chamber, the female will often dig out a depression and sit somewhat curled up for a few weeks or more before laying eggs.

Uncovered brood chamber.
© David T. Anderson

Scolopendra heros heros guarding her eggs.
© David T. Anderson

smaller species). Previous lack of knowledge about long-term sperm storage and post-molt sperm retention led to this strange theory.

EGG LAYING AND BROODING

In most areas, Scolopendromorpha produce one batch of eggs a year. They emerge from the earth in early spring in temperate climates and following monsoons in the deserts and tropical areas (with a dry season). Females spend a few months feeding to replenish their energy reserves, mate and then settle down to produce young. The time spent feeding is very important because they will spend approximately two months in the brood chamber without food while caring for eggs, protonymphs, and 1st instars. In the U.S., *Scolopendra viridis* and *Hemiscolopendra marginata* form brood chambers in late June, while *S. heros* and *S. polymorpha* usually begin around mid-July. Although seasonal variations and differences across geographic areas will alter the exact timing, females in the same area are found in similar stages of brooding. If the eggs are damaged or consumed, the female will not lay more eggs for many months, usually a full year.

There are commonly twenty to eighty eggs in a brood with an absolute maximum of 156 reported. Eggs are round, yellow to orange, and range in size from a pinhead to a large pea. The size of the egg is related to the species, not to the size of the female (within reason). A particularly small female of a species will lay fewer rather than smaller eggs. Rarely a large female will lay a small number of eggs but the eggs are of the standard size.

While the vast majority of centipedes abandon their eggs in cracks or bury them in the soil or beneath leaf litter and logs, all of the Scolopendromorpha practice some degree of maternal care. The mother excavates a burrow that is much wider but shorter than her body.

Protonymphs, *Scolopendra heros castaneiceps*. © David T. Anderson

First instars, *Scolopendra heros castaneieps*. © David T. Anderson

She evens out the interior of the cell but does not go to great lengths—the walls are rough. She fills in the entryway with dirt; she will not be leaving for a long time. Smaller Scolopendromorphs like *Scolopendra viridis* and *Hemiscolopendra marginata* usually construct the brood chamber under loose bark of rotten logs (about a foot above ground). Larger species like *S. heros* construct the chamber in the ground or beneath large logs or rocks. However, there is often no sign of a brood chamber in captivity because the substrate is shallow, dry, or cannot be easily formed, the shelter is too small, etc.

Ethmostigmus sp., Tanzanian "Giant Yellowleg" eggs, first instars, second instars (protonymph stage not pictured). © Kelly Swift

Brood chambers in nature possess at least five properties different from those in captive situations.

High humidity—even in the desert, brood chambers are constructed following seasonal rains and below ground. Humidity is crucial because the permeable eggs will not survive dryness. Also, many soils are too hard or loose for construction of the chamber if dry.

Minimal ventilation—excessive ventilation may not cause a problem unless the surrounding air is dry.

No light—the eggs of nearly every captive hatching are exposed to indirect room lighting at some point when the keeper finds or checks on them. Direct sunlight or incandescent should be avoided.

Consistent and often low temperature—even in the deserts and tropics, the temperature does not change much where brood chambers are formed. The temperature is often around 66°-70° F (16°-21° C) by depth even if the surrounding daytime temperatures are much higher. Captive broods are almost always kept at room temperature, around 75° F (24° C), sometimes much higher. But, centipedes do not lay their eggs in the sun, on top of a rock, and so they will not experience the temperature fluctuations exposed eggs would.

Minimal disturbances—the earth, rock or log has been there for years and is unlikely to move without an earthquake or bulldozer. Disturbances (including noise) are common in captivity. Centipedes do not have ears but they have numerous sensory setae to detect vibrations (if yelled at, the centipede will usually wince or go running around the cage—it's not a question of air disturbance since the reaction still occurs in an unvented cage). Giant centipedes are excellent mothers and might not eat healthy eggs, but it is important to avoid disturbances.

When a female is found with eggs it is important to keep the humidity high, cover the cage with a sheet, and place the cage in a cool, quiet area. Even when fertile eggs are left with the mother, there is usually a fifty/fifty chance she will eat them.

Years ago, on a camping trip, I lucked upon nine adult female *Scolopendra viridis* inside their brood chambers; each wrapped around a cluster of thirty to forty eggs. All the females were found in an old wooded area 1,000 ft by 20 ft. I collected two of the females. I considered leaving the egg clusters behind but then they would certainly die without their mothers. I was aware that even minor disturbances could cause females to eat eggs. All I had were two plastic bags. I scooped up the females with the eggs and put each in a bag. The females ran from the eggs between capture and putting them in the bags. The eggs were extremely fragile and a few popped—seemingly just from touching the sides of the bag. A few hours later, both females had recollected their eggs. The bags were moved a number of times to keep them out of the heat over the next two days and then spent about sixteen hours in the car on the trip home. At home, both females with eggs were placed in a plastic shoebox on a few inches of moist soil. The eggs hatched after a few weeks and three weeks later went through their first molt. I was surprised that despite massive disturbances the females did not eat the eggs.

Other times I have had great luck, such as with a female *Alipes* who had three consecutive broods and was checked on and photographed repeatedly with no losses. Our native *Scolopendra* are a bit pickier but can ignore some disturbances. Usually if any female is disturbed when first laying the eggs she will eat them all immediately. The danger tends to lessen with development. Egg eating is not about fertility—I dug up an *Ethmostigmus*

Alipes grandidieri eggs.

Alipes grandidieri protonymphs.

First instar *A. grandidieri*.

Second instar *A. grandidieri*, third consecutive brood.

trigonopodus holding first instar larvae and she proceeded to eat them as I watched in horror.

Early Development

Ova are like tiny water balloons, if water balloons did not have a stretchy casing. Motherly protection and care is essential to their survival. The female pays great attention to licking and cleaning the eggs with her palps and mandibles and rotates them using her maxillipeds (talk about a sledgehammer to pound in a nail). If eggs are taken from the mother they mold or develop improperly and without her their chances of survival are near zero even with the best attempts of human intervention.

Early development can be difficult to observe even if disturbing the mother were not a concern. So far, every female of every Scolopendromorpha species whose early development has been observed, in nature and captivity, holds the eggs tightly within their legs. Although some can run while holding the eggs, most species fold their bodies over or wrap their bodies around the eggs and drop many of the eggs if they are forced to flee. Some of the

Scolopendra heros and *Scolopendra polymorpha* take encircling the eggs a step further and can completely envelope the eggs and protonymphs so that not one is visible. Since they are already protected within a sealed brood chamber, this may provide some added protection against desiccation and seems to correlate with animals from drier areas of the range.

Prior to hatching the egg changes shape as the developing animal matures. This first stage known as the protonymph (or second embryonic instar) almost looks like a hatchling (or lightly segmented tube). On close inspection the legs are not separate and a membrane is still visible. This additional stage is only seen in the scolopendromorphs and geophilomorphs. The protonymph stage is also seen in all milliped orders and is essentially the same as the postembryo in certain arachnid orders like the spiders and whipspiders. After a total of three to four weeks (mostly spent as an egg, the protonymph stage can last less than a few days) the protonymph hatches to become a 1st instar centipede. At this stage the centipede is still very placid and pale, but able to walk. First instars resemble the mother but they are white in color and often have a dumpy shape and strange-looking head. They remain in the mother's gentle embrace and do not eat. They can survive without the mother's care at this time but if removed, most or all are likely to die.

In another three or four weeks the 1st instars will molt to 2nd instar and shortly after are ready to disperse and hunt. They gain color following this molt and, in some species begin with a very different color pattern than the adult. In a few days to a week they start to leave the protective grasp of the mother and can be found all over the cage, often in groups. The mother does not begin eating them and they do not become cannibalistic for a few weeks or longer, but this is the time to separate them. Cannibalism often becomes intense beginning around week three.

Scolopendra gigantea with eggs. © David T. Anderson

A Mother's Ultimate Sacrifice

Generally within the Scolopendromorpha, the female lives one to three years after producing young and continues to lay new batches of eggs each year, or at shorter intervals. However, female *Scolopendra galapagoensis* and *Scolopendra gigantea* experience a very different outcome. Like other members of the order, if the female is disturbed during or after egg laying she tends to eat her offspring. Then, many months to a year later she will be able to try again. If successful, she will brood the eggs through the protonymph stage to first instar. This is where the narrative changes.

Her offspring now consume her body while she is still perfectly healthy. When she is eaten she remains capable of living for years—she is not the victim of old age. At any moment the mother could choose to eat her soft-bodied babies before they have a chance to turn on her, but she lovingly allows them to consume her alive. They often leave only the hardened ends of their mother's maxillipeds behind. Although her offspring look rather weak and sport only a thin,

Larval *Scolopendra gigantea* have a look only a mother could love. Unfortunately not *their* mother, they ate her. © David T. Anderson

Close-up of the *Scolopendra gigantea* madibular teeth with first maxillae removed (exuvium). © David T. Anderson

Larval *Scolopendra gigantea*, the mandibular teeth stick out past the first maxillae. © David T. Anderson

Close-up of the *Scolopendra gigantea* baby face. © David T. Anderson

membranous exocuticle, they are hiding a deadly secret. At this stage they have strange-looking mandibles with massive "monster" teeth used for this special task. Still, even the most fiendish fictional monsters might hesitate to eat their own mother alive. (C. Schwarz warning: No anthropomorphisms here please. Animals cannot be judged by human standards.) (O.M.: I am judging them by fictional movie monster standards.)

At this point in time, such behavior and physical adaptation of the larval mandibles is only known for the *S. gigantea–S. galapagoensis* complex, but considering the large number of different Scolopendromorpha hatched in captivity so far, there is no reason to believe this adaptation is mirrored in other species. As a result, at dispersal the young are much larger than any other species. The selective pressure for size must be great indeed for these South American monsters.

Rearing Young

All species are kept separately for best survival—it is possible a few species could be

Larval *Scolopendra gigantea* molt as the body and teeth take on the normal appearance. © David T. Anderson

reared communally but the vast majority kill each other. Five-ounce containers with snap on lids easily contain the young for at least a few molts, and six to twelve months, unless it is *Scolopendra gigantea*. Rearing containers that are too large make it easy for food to go unnoticed. The lid should have half a dozen pinholes for ventilation. Small centipedes are especially prone to drying up, so there needs to be half an inch of moist—not dripping—substrate. Good substrate can be potting soil, peat or coconut fiber mixed with a little sand. A layer of loose shredded leaves or paper towel should be placed on top of the substrate. It is crucial the lid fits well because the tiny young dry up and die rapidly if they get loose. A large number of small containers can be placed inside larger shoeboxes or plastic tubs to guard against escape and drying out of the substrate. This also makes feeding easier since the group can be placed in the middle of the floor or near the feeder cage rather than trying to move dozens of small containers.

Live fruit flies, small isopods, worms, and freshly killed crickets or mealworms are possible food items for the immatures. Fruit or fruit flies can be beneficial to get feeding started, but should never be used as a primary or singular diet. Large live crickets can jump to evade capture for days or weeks, but are immediately consumed when offered dead. Live mealworms sometimes attack small centipedes, especially if they are nearly as big as the centipede. Only as many prey as will be eaten in a day or two should be placed in the container. In large numbers, even the timid

Scolopendra subspinipes mutilans mother losing interest in her second instar brood. © Kelly Swift

isopods could possibly stress or damage a molting centipede. A tiny piece of banana placed in the cage of the first molt centipedes (in addition to standard fare) seems to improve the survival rate of some species even if already feeding well.

The die-off rate has been very low for the majority of species kept in captivity (notwithstanding zero effort). For most scolopendromorphs loss of greater than 10% is due to cannibalism, improper prey, underfeeding,

Alipes sp. can run while holding the first instar young.

excessive warmth, cold, or other easily identified problem. Specimens that have dead body sections (two or more nonfunctional legs) are dying and will not recover.

With close attention, young scolopendromorphs can molt every month, though once every two to three months is common under normal rearing conditions. Within a few months the immatures may already be over 2″ (5 cm). The growth of sibling groups is somewhat uneven but if animals from the same brood are more than one or two molts apart the cause is almost certainly nutritional. Any that have grown large enough and strong enough to bite through skin should be transferred to a container with screw-top lid. There can be uneven growth of immatures, but usually they are only a molt apart when reared under the same circumstances.

SELECTED SPECIES OF SCOLOPENDRIDAE

New enthusiasts almost always ask which species is best to begin with. The various pet species are similar in requirements and hardiness but not longevity or price. *Scolopendra polymorpha* and *Scolopendra alternans* (Florida specimens) could be considered beginner species because they are easy to find, relatively inexpensive, and moderately long-lived if provided a minimum of care. However, extremely small immatures of these species are not suggested for beginners. Temperament is another consideration, but those that seem more passive may be more risky through lack of concern than those which warn of their disinterest in being molested.

For most species, it is best to start with an immature, between third and sixth instar for the best longevity. Second instars are easy to raise but often ship poorly and may have never eaten before purchase. Small immatures can be very hardy but early instars of a few species experience a high die off rate even when kept properly. Wild-caught adults tend to be near the end of their lives, though native species, very rarely imports, can live a few years or more. Tropical, imported, mature animals often die two to six months after purchase and it is impossible to tell if the animal died from old age or improper care without a large sample group and extensive experience. Immatures live longer and more quickly regenerate missing legs because they molt more often. A missing leg or two is usually not a problem, but avoid animals with broken antennae or body damage because they may never completely heal.

Species Notes Terminology

Indications. Some anatomical details such as the presence or absence of a ring furrow are inviolate but most are not so concrete. Antenna segmentation is an indicator but is not consistent for every individual of a species. Furthermore, in some species such as *Scolopendra alternans* and *Scolopendra gigantea* the number of smooth segments can change as the animal grows (small immatures may have more or less smooth segments than the same adult). Spines on the terminal legs can also be quite variable. Despite natural variation, if your specimen is missing structures as outlined here, it may well be misidentified.

Length. Normal adult length (head to tail) does not include antennae or legs. Here we will try to stick with actual measurements of real animals. There is a strong propensity to "eyeball" centipedes at many inches greater than their actual length since it is not always easy to directly measure an animal that can change its length as it moves and can cause the measurer extreme pain. Stated measurements are often much like the 49′ snake at a zoo in Java, Indonesia, later proven to be less than 23′ in length (the keepers were astounded the snake shrunk so much). A lot of estimates throw in the rear legs and sometimes the antennae for good measure. The majority of 4″ *Scolopendra polymorpha* seem to be labeled as 6″ specimens but certain individuals from specific localities can grow even larger (or smaller).

Coloration. Common color morphs can be widely variable but are often consistent for subspecies or geographical location. This refers only to the top-view or dorsal appearance.

Locality. Known collection, range, or export locations.

The habits of different species are discussed in relation to other centipedes. No healthy centipede can be truly described as slow or docile.

FAMILY SCOLOPENDRIDAE

Indications: Head capsule with 4 pairs of eyes.

SUBFAMILY SCOLOPENDRINAE:

Indications: Spiracle valvular, in shape of horizontal isosceles triangle.

Scolopendra alternans Leach, 1813
Indications: Antenna composed of 17 segments, first 5 smooth. Front legs have two claws while the terminal prefemora have a large number (+/- 20) of small spines mostly facing inward. Pair of faint grooves on head plate, only easily visible on dried specimens. Length (FL): 4-6.5″ (10-16.5 cm). Coloration (FL): Overall yellowish to golden brown. *Scolopendra alternans* from Florida is one of the few giant centipedes which tends to be monochromatic, but the tips of the antennae or legs can be a little greenish. Locality: Dade and Collier counties in southern Florida and Keys, also Antigua, Bahamas, Barbuda, Brazil, Cuba, Dominica, Guadeloupe, Haiti, Montserrat, Puerto Rico, St. Barthelemy,

Scolopendra alternans adult from Florida. The antennae are half green on this specimen.

Two-year-old, captive-hatched *Scolopendra alternans* Haitian.

Four-year-old, captive-hatched *Scolopendra alternans* Haitian adult.

St. Lucia, St. Kitts, Venezuela, and U.S. Virgin Islands (Shelley 2002).

Scolopendra alternans is a nervous, shy centipede that prefers to keep its body buried with just the rear legs or antennae sticking out of the substrate. It stays buried when the cage is moved but torpedoes out of the dirt when touched by prey or forceps. *Scolopendra alternans* Florida is not as readily available as some imports but it is not difficult to locate and is relatively inexpensive.

Scolopendra alternans Haitian adult.

Scolopendra alternans
"Puerto Rican Giant" or "Haitian Giant"
Indications: Same as for FL *S. alternans.* Length: 4-8" (10-20 cm) Coloration: Body greenish-yellow with dark bands along the rear of tergites 1-20, antennae and legs fade from orange into green. Locality: Puerto Rico, Haiti.

There are no physical differences that would easily differentiate this centipede from *S. alternans*, so it is almost certainly just the very colorful Caribbean form of this species. Likewise, it is shy and skittish. Broods contain from 50-60 tiny eggs. The 2nd instars are emerald green. The tiny young are somewhat sensitive and can experience as high as 30% die off during the first three or four instars despite adequate care. Afterwards, the die off should be close to zero.

This centipede is especially shy and reclusive, and is usually found deeply buried in the substrate. When irritated, it wildly flops around like a fish out of water. The 2nd instars are relatively tiny and can take four years or more to reach maturity. *Scolopendra alternans* grow more slowly than other large species because they start out maybe a third the size of *S. heros* or *S. subspinipes* and require more molts to reach maturity.

Haitian "Black" 3rd or fourth instar.

Scolopendra alternans
"Haitian Black"
Indications: Same as for *S. alternans*. Coloration: Variable in the same brood.

The original imported specimens were nearly solid black. One of the smallest, a 5" (13 cm) female, produced fifty-two young after nearly a year in captivity. All the immatures were dark from second to fourth instar. However, these turned out not to be an isolated color form, but simply one coloration of the "Haitian giant." These captive-hatched young from the black female were reared to adults and turned out to have not only the same morphology, but also, the majority ended up with the same color pattern as the "Haitian giant." At first they seemed to be something different because the early instar immatures had one additional smooth segment (the wild-caught mother was no longer available to check at the time). However, the number of smooth segments changed back to the standard five as they grew larger. Since this has not been tracked in other *S. alternans* there is no way to know if

Haitian "Black" large immatures mostly reverted to standard color though they were very dark in the early instars.

Puerto Rican adult *Scolopendra alternans*.

Haitian "Giant" third or fourth instar.

the antennomere transformation is normal for all members of the species, only those from a certain geographic area, or even a single brood. Furthermore, the development of lighter coloration in the adults may be standard (possibly only a small percent remain black in nature) or the father may have been the lighter banded form. It is also quite possible dietary deficiency in captivity did not allow for the formation of enough dark pigment since all the young were a dark smoky black in the early instars.

Recently some large reddish specimens from Hispaniola (the island containing Haiti and the Dominican Republic) have shown up in a few rare imports. Despite one or two morphologically impossible identifications being bounced around, these appear to be a color variety of the "Haitian Giant" *S. alternans* from a new collection location. Most endemic Neotropical *Scolopendra* have a ring furrow, with *S. alternans* an obvious exception. The only other large species, that is also well known from Hispaniola, the Old World introduced

Scolopendra alternans "Hispaniola," five smooth antennomeres and no ring furrow. © Amy and Gabriel Tendera

Scolopendra alternans "Hispaniola," with terminal prefemoral spine formation, standard for this species. © Amy and Gabriel Tendera

Scolopendra alternans "Hispaniola," size reference. © Amy and Gabriel Tendera

Scolopendra angulata. © John Skierski

Scolopendra subspinipes, also lacks a ring furrow but it does not have the large number of small spines on the inside of the terminal prefemur, and it has six rather than five smooth antennomeres.

Scolopendra angulata Newport, 1844
Indications: Antenna composed of 17 segments, first 4 smooth. Ring furrow present. Rear legs with small spines on femur. Tergites 20 and 21 have particularly flared edges. Length: 4-6″ (10-15 cm). Locality: Guyana, Suriname, Barbados.

This species is rarely seen in collections but shows up every so often as wild-caught adults, probably never as captive-hatched. Specimens can be even fatter bodied than *S. polymorpha*. *Scolopendra angulata* never seem to refuse prey, even when obviously overfed or approaching a molt. The preference is for very moist conditions (think coffee grounds). They hide almost as much as *S. alternans*, usually under the water dish in a dug out, circular burrow. Photographs do not seem to pick up the iridescent quality of the tergites. They are slow and lumbering compared to most other *Scolopendra*. There are few longevity reports, but two imports lived for the better part of two years. Each molted once and expired a few months after the molt (J. Skierski pers. comm.).

Scolopendra angulata © John Skierski

Scolopendra angulata terminals have multiple prefemoral spines, as well as femoral spines similar to *Scolopendra hermosa*. © John Skierski

Scolopendra cingulata. © Frank Foldbjerg Hansen

Scolopendra cingulata Latreille, 1829
Indications: Antenna composed of 19 or 20 segments, first 6 smooth. Ring furrow absent. Length: 3-4.5″ (7.5-11.5 cm). Coloration: Tan body with black tergal margins thickened toward middle of tergite joints, rear legs orange (color for specimens from Spain and southern France). Those found elsewhere can range from nearly white to solid black. Locality: Southern Europe to North Africa.

The "Mediterranean banded" centipede has been billed as a hardy, medium-size species, but the few specimens I have kept barely lived a year (admittedly hardiness and longevity are not the same thing) and hardly exceeded the lowest level of the length spectrum. The short captive lifespan could easily relate to the age of the animals or result from improper care in import facilities. *Scolopendra cingulata* is commonly kept in Europe but seldom seen in the U.S. hobby. Coloration is variable across the geographic range and it often resembles a small, banded *S. polymorpha*.

Scolopendra dehaani Brandt, 1840
Indications: Antenna composed of 16-18 segments, inner 6-8 smooth. Ring furrow absent. Head lightly punctuated. Rear legs with two or three inward facing spines on the terminal prefemur, none on the underside. Length: 5-7.5″ (13-19 cm). Coloration: Reddish-brown body with yellow legs, except the last three leg pairs orange, sometimes all orange with darker terminals like some of the Chinese *Scolopendra*

Scolopendra cingulata collected in Italy. This specimen must have lost and regenerated the smaller terminal in the wild. © Frank Foldbjerg Hansen

Scolopendra cingulata white form, possibly from Egypt.

Scolopendra cingulata from Zakynthos, Greece. © Frank Foldbjerg Hansen

Scolopendra dehaani, the common but still very handsome Vietnamese Yellowleg.

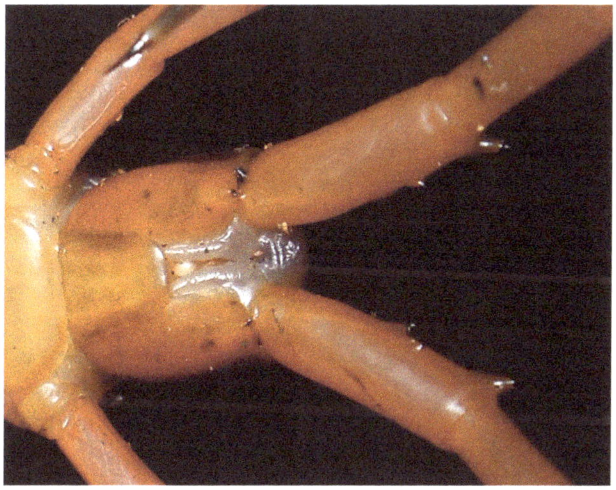

Scolopendra dehaani orange leg terminals from below, no spines.

Vietnamese Yellowleg terminals.

subspinipes. Locality: India, Burma, Bangladesh, Malay Peninsula, Andaman Islands, Nicobar Islands, China, Hong Kong, Indonesia, Thailand, Vietnam, Japan, Okinawa, Cambodia (Kronmüller 2012).

Scolopendra dehaani had been reduced to a subspecies of *S. subspinipes* for a time before being returned to full species status, so it is often labeled as *Scolopendra subspinipes dehaani*.

Vietnamese "Yellowleg": This has been one of the most popular species for years simply because specimens are large and handsome, but mostly because they are cheap and readily obtained through wholesale. The imports from Vietnam tend to be less "grumpy" than large *S. subspinipes*. They usually arrive full-grown, but undamaged specimens still continue to live a year or two, sometimes longer after import. The number of reproduction reports is quite tiny when compared to the number of imported specimens. The listed common name, colors, and experience of course only refers to the rather consistent imports from Vietnam, but sometimes the legs are orange.

Scolopendra dehaani from Malaysia, "Cherry Red" adult. © Lee Chia How

The fantastic "Cherry Red" 2nd instar. © Lee Chia How

"Cherry Red," after a few molts the color grows closer to the adult. © Lee Chia How

Scolopendra dehaani. This specimen was a stow-away inside a box coming from Indonesia which was supposed to only contain an aircraft motor. It was winter 2006 and she nearly died from the cold. (The airport called Tom looking for help since he is well-known for invertebrates and airport personnel had no interest in keeping the centipede as a pet.) This female *S. dehaani* lived a little over a year, molted one time, and laid a batch of eggs which she later ate. © Tom Larsen

The same Indonesian *Scolopendra dehaani* after laying eggs. © Tom Larsen

Malaysian "Beauty," same specimen with a pale, regenerated leg. © Tom Larsen

Malaysian "Cherry Red": The uncommonly imported adults from Malaysia are a beautiful reddish brown with dark banding. One of the 2001 imports, a six-inch (15 cm) female produced sixty-five young, but at the time the color of the 2nd instars was a great surprise. They were light orange with black banding and had bright blue heads, antennae, and first two tergites, including the attached legs (Ditzel 2002). After a few molts the heads change to gray and the body darkens.

Others: There are quite a few color patterns including "Myanmar flamelegs" from Thailand. Supposedly the spectacular Malaysian "beauty" is a geographic color race of this species.

Scolopendra galapogoensis Bollman, 1889
Indications: First 4-7 antenna segments are smooth. Ring furrow present. Most legs have 3-5 small black spines on the dorsal apex of the prefemora. Only the first pair of legs, usually, has a femoral spur. Two grooves run along each sternite, front to back, and divide the area into three roughly equal sections. Length 6-8″ (16-20 cm). Locality: Galapagos and Cocos Islands, Pacific Coasts of Peru and Ecuador, west of the Andes.

In the giant centipede handbook, *S. galapagoensis* was illustrated only for differences from *S. gigantea*. This species for years had been considered very uncommon, possibly never seen in captivity, but that may have been because specimens were almost always labeled *Scolopendra gigantea*. Specimens have also been incorrectly listed as *S. gigantea weyrauchi*, a synonym of *S. galapagoensis*, or even *S. gigantea* "robusta" even though *Scolopendra robusta* is actually an entirely different, small Mexican species. Some of the *S. viridicornis* and *S. gigantea* imports have been labeled *S. galapagoensis* making it difficult to judge just how rare this species has been in captivity. Ultimately, like *S. gigantea*, it is a massive species and at 7″ (18 cm) it is nearly double the bulk of a similar length *S. heros*. Over its range it seems to be similar to *S. gigantea* in its range of colors, since specimens can be tan to orange or dark with banded legs. These two species may be a species complex since they are similar in morphology and the differences are not so distinct in many specimens because of variation in the femoral spurs and smooth antennomeres. Furthermore, these are the only two species where the larvae are strangely adapted for eating their own living mother. The tips of the long, sharp teeth stick out between the first maxillae which is unusual indeed (mandibles are normally well hidden behind the first maxillae) but they return to normal size at the next molt.

Scolopendra dehaani “Myanmar Flamelegs” from Thailand. © Chia How

Scolopendra gigantea? Venezuela, 2004. © Tom Larsen

Scolopendra gigantea Linnaeus, 1758

Indications: Antenna composed of 17 segments, first 7-10 smooth. Ring furrow present. Most legs have 3-5 small black spines on the dorsal apex of the prefemora. Many legs have femoral spurs. Two grooves run along each sternite, front to back, and divide the area into three roughly equal sections. Length: 7-10.5" (18-27 cm). Coloration: Red brown to dark body, brown or cream legs, banded in the dark form. Locality: Aruba, Curacao, Isla Margarita, Colombia, Venezuela, Suriname, and southern Panama.

Scolopendra gigantea was the first named scolopendromorph and its story underlines some of the perplexing aspects of taxonomy which are not limited to this species. The original description of a specimen from Haiti (Linnaeus 1758) included drawings of the prefemoral and femoral spines but there was no type specimen until 2000 (Shelley and Kiser 2000). The original drawing for *S. gigantea* is missing two body segments, attributed to error since no Scolopendromorpha has 19 segments and specimens have never been found in Haiti, only the countries listed above. Moreover, some populations are missing femoral spurs on multiple legs which combined with variations in the smooth antennomere count means it is not always possible to definitively differentiate the two species without collection location. Real animals do not always fit perfectly within species concepts. According to distribution, if listed correctly, the two seem to be vicariant species, with *S. galapagoensis* west and *S. gigantea* east of the Andes. Morphological separation may not be severe when species ranges do not overlap and selection does not cause morphology to divert. Some

This female was only about seven or eight inches (less than 20 cm) when she first laid eggs and ate them. She grew larger at the next molt, *Scolopendra gigantea*. © David T. Anderson

Scolopendra gigantea have three to five small spines at the end of the prefemora and one small spur or spine at the end of the femora of most legs (located dorsally).

Adult *Scolopendra gigantea*. © David T. Anderson.

Adults of the same bloodline eventually have seven or more smooth antennomeres (the maximum is uncertain).

This youngster has just four smooth antennomeres. © David T. Anderson

"Where did mom go guys?" © David T. Anderson

Scolopendra gigantea large brood. © David T. Anderson

Peruvian specimens were placed in the subspecies *S. gigantea weyrauchi* (Bucherl, 1950) but the holotype is lost and is presently thought to be *S. galapagoensis*. Furthermore, as depicted, smooth antennomere count can change with age.

A few color forms are known for this species—those brought in during the early 1990s were chestnut brown with lighter legs. Specimens brought in the early 2000s from Venezuela looked similar in shape but the bodies were more reddish brown and the legs a brighter white. The last color form is dark brownish black with banded legs and looks similar to the dark *S. galapagoensis*.

This centipede is skittish, powerful, and the largest centipede in the world. It prefers to stay buried with the antennae and rear legs jutting from the surface. A large specimen can consume an entire pinkie or large *Blaberus* nymph without leaving scraps. Specimens often lose their appetites below 75° F. This species produces a large number of eggs, up to 156 in a brood (D. T. Anderson pers. comm.). Unlike the majority of centipedes, a female that lays eggs

Scolopendra hiananum adult male. © David T. Anderson

(and eats them of course) can gain measurable size (though not significant) in a subsequent molt. Post-sexual maturity growth, while rare for centipedes, is seen in a few milliped species, arachnids that have postultimate molts, and nearly all crustaceans (McMonigle 2011, 2012).

Scolopendra hainanum Kronmüller, 2012
TIGER LEGS CENTIPEDE
Indications: Antenna composed of 17-19 segments, first 6 smooth. Ring furrow absent. Heart-shaped genital sternite. Length: 6-8″ (16-20 cm). Coloration: Dark olive trunk with reddish head and first tergite. Yellow legs with dark bands on the female. Locality: Hainan Island, China

This is a nice, large, handsome species with dimorphic adult coloration. Tiger legs have been available numerous times as captive-hatched young in recent years. This popular species was, until 2012, thought to be a race of *Scolopendra subspinipes dehaani.*

Scolopendra hardwickei Newport, 1844
INDIAN TIGER CENTIPEDE
Indications: Antenna composed of 18 segments, first 6 smooth. Ring furrow absent. Length: 5-7.5″ (13-19 cm). Coloration: Orange and black alternating solidly colored tergites. Locality: Ceylon, India, Java, Nicobar Islands, Singapore, Sri Lanka, Sumatra (Lewis 1981).

This is arguably the most beautiful centipede in the world. The stark alternating black and orange tergal pattern is amazing. Second instar *S. hardwickei* start out colorful, with the same general color and pattern as the adults. However, the orange tends to be brighter and more intense on small animals. Photos that started showing up around 2006 were thought to be fakes by some. This was enhanced by the fact there are a number of strange color aberrations which resemble human error. There are

Scolopendra hainanum adult female. © David T. Anderson

Scolopendra hardwickei often look mispainted. © David T. Anderson

Older adult *Scolopendra hardwickei* often lose the bright colors of youth. © David T. Anderson

often splotches of color spilled over from one segment to another or corners that appear to have missed a brush stroke. The back-to-back orange tergites halfway through also seem like a mistake.

The second instars start out large but so far they have proven rather touchy in captivity with huge losses (50% or better over months or years) experienced by various keepers. Some keepers have reported immature *S. hardwickei* losing function of their legs over large sections of the body. This ultimately always results in death and may just be a symptom of dying animals rather than a disease.

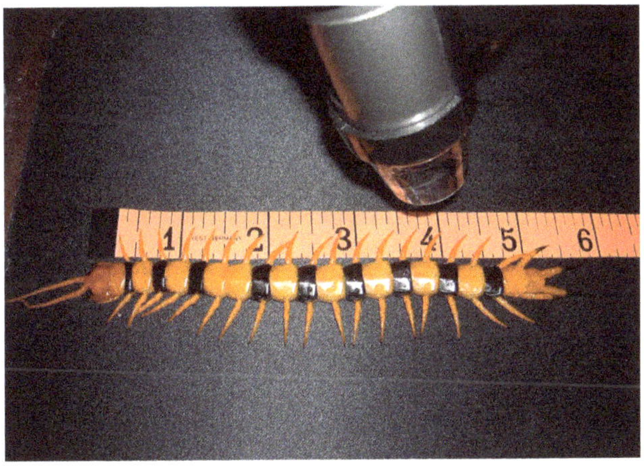

Scolopendra hardwickei trunk length (14 cm). © D. T. Anderson

Scolopendra hardwickei immatures. © David T. Anderson

Scolopendra hardwickei appear rather jerky and mechanical in their movements when they are small. This unusual form of movement goes away as the centipedes get older and larger.

These should be kept warm around 80° F for the best growth but warm temperatures have not proven to be the deciding factor. They mature in two to three years with proper care but may take much longer if underfed or kept cool (Anderson and McMonigle 2013). Although rarely imported except as immatures, it seems a large percentage of wild females produce young in captivity.

Scolopendra hermosa Chamberlin, 1941
Indications: Antenna with 17 segments, first 4 antennomeres smooth. Ring furrow present on first tergite. Terminal prefemur with around a dozen spines on the inner surface and also five or so inward facing spines on the terminal femur. Length: 5-6" (13-15 cm). Coloration: Greenish brown body, head plate orange-yellow towards the rear and maxillipeds of the same color, legs yellowish. Locality: Peru.

A number of these came in from Peru in 2009 and lived a year or so. It is a very handsome, heavy-bodied species with a unique look.

Scolopendra hermosa. © John Skierski

Scolopendra heros Girard, 1853
Indications (all subspecies): Antenna composed of 26 or more segments, first 3-5 smooth. Head finely punctuated with two faint grooves. Palps have a notable inner tooth. First tergite with ring furrow and fine punctation. First pair of legs outfitted with two claws. Length: 5-7.5″ (13-19 cm).

Scolopendra heros is slower moving and more docile than many species which grow as large. Photos of enthusiasts handling a huge centipede tend to be this species since it is usually "calm" so human hands can keep up with it. Specimens like to sit under flat objects but almost never burrow themselves in the substrate. Despite monstrous size, they are seldom enthusiastic about foods larger than an adult cricket. Broods usually contain thirty to fifty eggs. Captive-hatched specimens take two to three years to reach sexual maturity. They can take longer but stunted, poorly cared for specimens are more likely to die than reach adulthood.

The occurrence of all subspecies of *Scolopendra heros* is usually associated with rocky areas. It is rarely found in east Texas—mostly in rocky outcrops that are not common in the region. Specimens are commonly found within walking distance of water sources. Arizona specimens are also found in rocky areas, often in ravines or mountainous areas rather than cactus deserts or sand flats.

Scolopendra heros arizonensis, spines on the terminal prefemoral process or spur. © John Skierski

Scolopendra heros arizonensis adult. © David T. Anderson

Scolopendra heros arizonensis large immature.

Scolopendra heros arizonensis unbanded from Nogales.

Scolopendra heros arizonensis unbanded.

Scolopendra heros arizonensis feeding. © John Skierski

Scolopendra heros is the largest North American centipede with three separate and spectacular color phases. Historically the three have been considered subspecies though the geographic ranges had not been well-defined (McMonigle and Anderson 2011a) despite being widely published. These variants or races were referred to by Shelley (2002) as "color pattern traditionally assigned to the variant named castaneiceps by Wood (1861): . . ." which seems a roundabout way of recognizing the uniqueness in an attempt to avoid it. The stark contrast in forms without a single documented intermediate wild form has brought some question that they might even prove to be very closely related species were DNA analysis ever performed. Captive crosses have resulted in abnormally colored animals that match no known wild specimens. Subspecies by definition should have natural intergrade forms where boundaries overlap. What keeps nearby populations from interbreeding in nature is unknown, but there are some natural boundaries.

Scolopendra heros castaneiceps adult female.

Scolopendra heros castaneiceps adult male.

One anecdotal report suggests mate selection is possible: "I have been attempting to mate a female *S. h. arizonensis* with a male hybrid, *S. h. heros* banded x *S. h. castaneiceps*. The male refuses the female's advances by tucking his terminals underneath his rear sternites each time, through three attempts, so far" (Skierski pers. comm.). Of course this does not explain what may make certain mating attempts possible.

Scolopendra heros arizonensis, mature unbanded specimen, next to ruler. © Wade Harrell

Scolopendra heros castaneiceps freshly molted. © David T. Anderson

Scolopendra heros Subspecies

Scolopendra heros arizonensis Kraepelin, 1903
Consistent coloration: Last three tergites (dorsal body segments) black, bright orange terminals and antennae, head dark, legs whitish tan.

Common variations: Two primary variations, unbanded ("normal") and banded. The first tergite is usually black on unbanded specimens, tan on banded. Body coloration ranges from tan to reddish orange, reddish orange coloration usually only on small, young specimens.

Range: Arizona except for northwestern and extreme western regions, southwestern "boot heel" of New Mexico (Animas and Peloncillo Mountains in Hidalgo County), USA.

Scolopendra heros castaneiceps Wood, 1861
Consistent coloration: Orange head and first two tergites, dark body, light orange legs though the terminal pair is the color of the body and transitions to orange towards the tarsi.

Common variations: Orange can vary from light orange to reddish orange. Adults are greenish to brownish black. The bodies of some older specimens fade to greenish brown. In many hundreds of second instar specimens seen in captivity, the main body starts out bluish black.

Range: Northeastern New Mexico, all of Texas except the southwest, east through western Arkansas and Louisiana, north to Kansas, and southwestern Missouri. Colorado is commonly listed and the southeastern tip follows the likely range, but there are no verifiable reports available. Attems (1930) includes Georgia and Alabama, but there are no verifiable reports.

Scolopendra heros heros Attems, 1930
Consistent coloration: Light, golden tan body, lighter legs, tan antennae, last two tergites are black.

Common variations: Banded specimens have a dark head. Unbanded, the common or "normal" variant has a light colored head. Terminal legs are usually solid tan on banded specimens but dark at the base, transitioning to tan on unbanded.

Range: Southwestern Texas (see map) to south central New Mexico north and east of "boot heel" region, also northwestern Mexico.

Scolopendra heros castaneiceps fourth instar, captive-hatched.

Scolopendra heros arizonensis, mature unbanded specimen collected in July of 2012 in Tucson, Arizona, and molted in June of 2013. © Wade Harrell

Minor banding and color variations within subspecies may cause some confusion, but overall color patterns still hold true. In addition, the exact shade of a color is influenced by the age of the animal. Furthermore, color depth, not pattern, can change when reared under captive conditions as a result of nutrition or environment.

Average-sized mature adult *Scolopendra heros heros* at 6 1/2″. 6.5″ trunk length is common for a large *Scolopendra heros heros*. It is a big animal—most people would "eyeball" this at nine or ten inches. © David T. Anderson

Blue-tinted terminals continue until animals are three or four inches long. *S. heros heros*. © David T. Anderson

There was a fourth variant described, *prismatica*, by Craigin in 1885, but there is no type specimen and nobody has seen one since 1885.

Heros in the Hobby

Availability of captive specimens seems to fairly reflect the size of the natural range. The

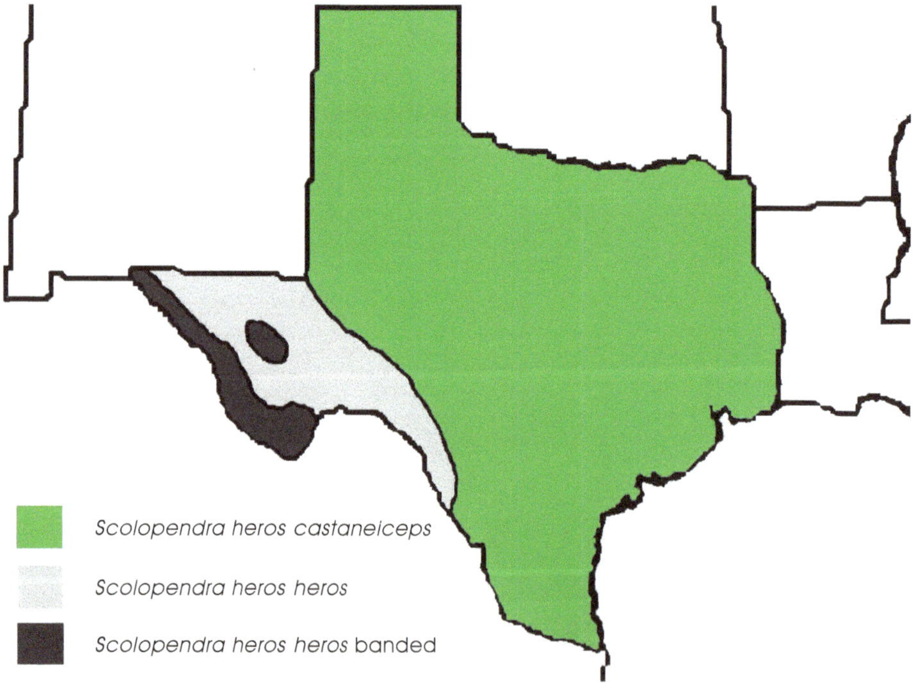

Scolopendra heros distribution in Texas. Reproduced from McMonigle and Anderson (2011a)

The coloration of this *Scolopendra heros heros*, third instar captive-hatched specimen explains the "blue tail" common name.

Scolopendra heros heros banded form.

most widespread, *S. h. castaneiceps* "giant red-headed centipede" is always available from one vendor or another. Banded and unbanded *S. h. arizonensis* "Arizona giant centipede" are regularly available in limited numbers. Specimens of normal *S. h. heros* "giant bluetailed centipede" are found in a much smaller range and are very difficult to acquire. The even more restricted banded form of *S. h. heros* is the rarest and only the most ardent enthusiast will ever see one in person.

The banded form of *S. h. arizonensis* was known and listed at least as far back as Attems (1930), but not that of *S. h. heros*. The banded *S. h. heros* is even rare in literature and may not have been formally reported prior to 2011 (McMonigle and Anderson). On the other hand,

Scolopendra heros heros older adult male.

Sandefer (1998a) has a picture of a banded *S. h. heros* specimen that must have come from Texas mislabeled as *S. h. arizonensis*—which does not follow the description in the same text—along with two photos of *S. h. arizonensis* correctly labeled.

Due to large size and bright colors, all varieties are among the most sought after centipedes

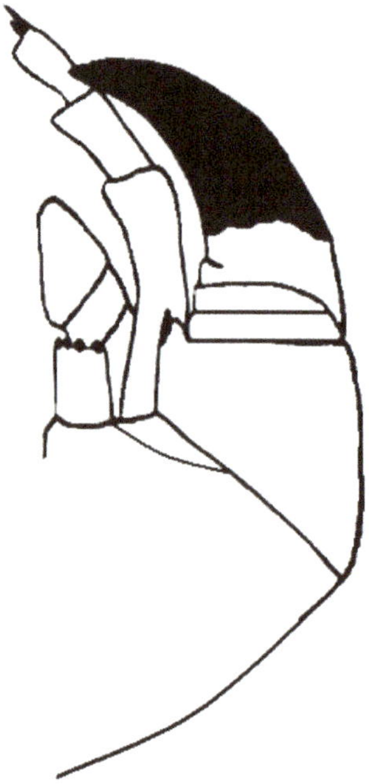

The second maxillary palps of *Scolopendra heros* are lamellate or toothed. This is seen in other species such as *Scolopendra gigantea* but never *Scolopendra polymorpha*.

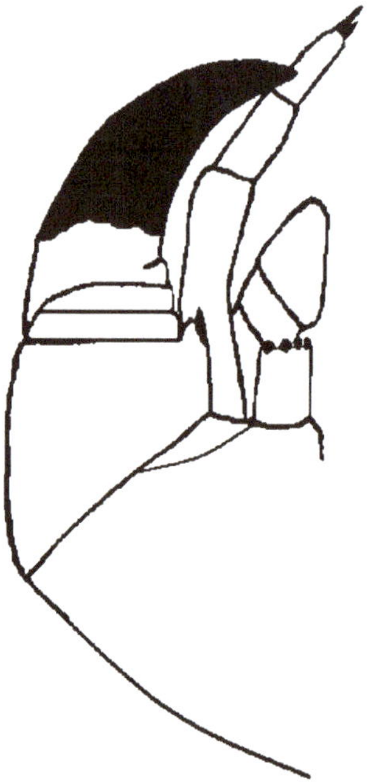

Scolopendra polymorpha maxillary palps do not have large, flattened inner teeth.

This specimen purchased as a "Golden *Scolopendra heros*" from southwest Texas is in fact *Scolopendra polymorpha*.

kept by enthusiasts. The banded form of *Scolopendra heros arizonensis* is the most coveted and commands the highest prices.

The Myth of the Golden *Heros*

The chance of finding a "golden *heros*" is as likely as finding the New World cities made of gold. The reports of solid yellow/gold *S. heros* from southwest Texas began only around 2006. The original specimens were "identified" by the collectors. The specimens I have seen collected and sold as "gold" *heros* were large yellow *S. polymorpha*. Whereas a big yellow *S. polymorpha* might fetch $25, a "golden *heros*" could fetch up to a few hundred dollars. I paid a good price for mine and was hoping to recoup much of the cost by reselling it. However, there was no question it was *S. polymorpha* from the palp shape (the number of sparsely hirsute antennomeres, smaller antennae, and body shape are other clues). I had to eat the full cost since I could not lie and call it *S. heros*, and nobody wanted to pay much for a big yellow *S. polymorpha*. I do not think the original collector purposefully falsified the "identification" but simply figured it was big enough to be *S. heros* and had no incentive to get an adequate identification. Some photos of a specimen made it to the website Nadiplochilo labeled *S. heros* but actual identification was again not attempted. Palp photos (not on the site) of the very same specimen easily prove it is *S. polymorpha*. The myth was more recently forwarded by Evans (2007) and unfortunately it is probably one of those myths that will grow rather than die.

The only species that ever seems to be confused with *S. heros* is *S. polymorpha*, but there

Scolopendra heros hybrid, third instar *castaneiceps* x *heros* banded.

Scolopendra heros hybrid, second instar *castaneiceps* x *heros* banded.

are a number of easily observed differences. A unique feature of *S. heros* is that the second maxillary palps are toothed rather than tapered, but this feature is rarely visible in photographs and can be difficult to see on live specimens. One feature is commonly visible in photographs and in person: *S. heros* usually have 3-5 smooth basal antennomeres whereas *S. polymorpha* normally have 7 or more. Although the number of antennomeres can be the same overall, the size of *S. polymorpha* antennae is often half that of *S. heros* because each antennomere is smaller.

The reason for the confusion is that both can grow large and are found in many of the same areas. While some *S. polymorpha* can be sexually mature at 1.75″ (4 cm), others are capable of exceeding 6″ (>15 cm) which is the average size of adult *S. heros* (which range from 5″ to 7.5″ (13-19 cm) at maturity).

Heros Hybridization

In 2009-2010 there were at least three successful captive crosses of *S. heros* subspecies found in Texas and Arizona. The results of these are fascinating but of course anecdotal. The appearance of the offspring is interesting in that there has been no major variation between siblings and while similar in pattern, coloration is different from any known wild specimens. The color patterns of hybrids shift as they mature, but both juvenile and adult states remain unique from any known wild specimens. It is of note that the second two crosses listed are between the same subspecies but banded and unbanded versions favored a different overall body pattern.

Scolopendra heros castaneiceps x *S. h. arizonensis* (unbanded): The parent subspecies are located the furthest apart. The resulting brood contained twenty individuals. Husbandry issues may or may not have been the culprit for extremely low survival, only one small specimen lived for a few years. Second instar animals resembled *S. heros arizonensis* in pattern but changed to all black, except for the legs and antennae, following a few molts.

Scolopendra heros hybrid, large *castaneiceps* x *heros* banded. © David T. Anderson

Scolopendra heros castaneiceps x *S. h. heros* (banded): The unbanded form of *heros* occurs between these two. Resulting offspring were in good number (thirty plus), initially robust, hardier and larger in early instars than normal *S. heros* immatures. Color pattern resembles *S. h. castaneiceps*, but the legs and tips are yellow rather than orange and there are

Scolopendra heros hybrid *arizonensis* x *castaneiceps*. (All hybrids pictured in this text were produced by adults artificially crossed in captivity; no wild hybrids are known.) © David T. Anderson

Large immature *Scolopendra heros heros* unbanded x *Scolopendra heros castaneiceps.*

Scolopendra heros heros banded x *Scolopendra heros castaneiceps.* © John Skierski

Scolopendra heros x *castaneiceps*
2nd generation. © David T. Anderson

Scolopendra heros x *castaneiceps*
2nd generation. © David T. Anderson

dark smudges on the head and front tergites. The dark smudges have grown larger and darker with each molt but so far not as dark as the preceding hybrid. The body is black but with a strong green hue rather than the blue cast of young *S. h. castaneiceps*.

Scolopendra heros castaneiceps x *S. h. heros* (unbanded): The natural range of these is right next to each other so it's interesting that the resulting cross had rather poor results—only eleven weak babies. Coloration of early instar young was similar to unbanded *S. h. heros*, but the main body grew greenish-brown instead of tan in older specimens and the head and first two tergites may be a slightly darker orange. The dark body color emanates from below the surface of the exoskeleton.

Conclusion

Whatever you want to call them, the three main color races of *Scolopendra heros* occurring in the United States are unique and will likely retain their names as long as there are people interested in observing or keeping these animals. If there were dozens of major color forms or they had nothing to do with geography and genetics or had no consistency within populations, subspecies designations would be uncalled for. However, so called "other forms" are standard variations within subspecies or misidentified *S. polymorpha*. Banding width or age-related variations within subspecies should never be given names.

Captive-reared hybrid specimens have ended up with unique shades and color patterns that changed more drastically during development than age-related shifts known to occur within subspecies between 2nd instar and maturity. Sibling coloration has been relatively consistent for first generation hybrids, with some reversion and variability seen in the second generation. Hybrids of *S. h. castaneiceps* x *S. h. heros* (unbanded) were reared and later mated back to each other. Captive crosses have resulted in interesting colorations unknown in the wild and add an interesting dimension to the subspecies discussion. Hybrids are ultra-unique specimens for the serious collector but are incredibly rare and certain hybrids may be plagued with health problems. Others display hybrid vigor.

Scolopendra heros heros with babies starting to color up.

Immature "Neon Blue" *Scolopendra mirabilis*.

Scolopendra mirabilis adult from Tanzania, 2001.

Scolopendra mirabilis (Porat, 1876)
"Neon Blue" or "*Scolopendra* sp. Tanzania"
Indications: Antenna composed of 18 segments, inner 6 smooth. Male without notable spurs on the terminal legs. Length: 3.5″ (9 cm). Coloration: Cream body with solid blue tergal margins and a wide blue band running up the middle of the tergites, bright turquoise blue legs, and a light orange head and first and last tergite (the orange is red and turquoise is blue on young animals). Adult males tend to be dark blue and females baby blue. Locality: Africa, India, and southwest Asia.

This is arguably the most brilliantly colored centipede available. Photographs do not do them justice. Some of our small *Scolopendra polymorpha* start out looking a little like the adult of this species when young (in pictures), but the juvenile color fades quickly and nothing else has such a brilliantly blue-striped body. The juvenile *S. mirabilis* has an entirely different range of colors when small. Though a good eater, specimens grow slowly. It usually sits on top of the substrate even if there are 4″ (10 cm) of workable substrate. They tend to be docile and rarely bite if touched (but certainly bite if picked up). The "Neon Blue" is difficult to find but well worth the search. (Part of the reason it is hard to find may be because dealers usually keep this one for themselves.) Captive hatches are rarely seen despite quite a few imports of mature females over the years.

Scolopendra morsitans Linnaeus, 1758
Indications: Without grooves on the head or coxosternum. Tergite 21 with middorsal

Scolopendra mirabilis from Tanzania. © Frank Foldbjerg Hansen

Scolopendra morsitans from Tanzania. © Frank Foldbjerg Hansen

Scolopendra morsitans. © John Skierski

groove. Sternites 2-20 have the common lengthwise paired grooves. Terminal prefemora with 9-10 ventral spines in three rows. Length: 3″ to 3 3/4″ (7.5-9 cm). Coloration: Tan, often with reddish or dark head, dark margins at rear of tergites, and pale to red tinged legs. Locality: Wide-ranging Old World species introduced to the West Indies, Mexico, and Central America, also Florida and Hawaii, USA (Shelley et al. 2005, Shelley et al. 2014).

Scolopendra morsitans is the type species for the genus *Scolopendra* though *S. gigantea* was described at the same time. This species is not often seen in the hobby since small size and somewhat standard coloration are rarely in demand. It is a rather widespread animal and imports, commonly from Tanzania, have shown up a few times being sold as "baby giants." These specimens are usually full grown.

This species has been intercepted in quarantined ornamental plant shipments from overseas a number of times, but one specimen was caught in Jacksonville, Florida, USA (Shelley et al. 2005). So far, no Florida collectors report finding this species.

Scolopendra sp.
"African Tiger"
Indications: Antenna composed of 20 segments, inner 6 smooth. Length: 2.5-3″ (6.5-7.5 cm). Coloration: Head and first tergite are dark green or red, tan body with black tergal margins, yellow legs, and orange terminal legs. Locality: Africa, Tanzanian imports.

This small, thin, elongate species has been occasionally common and inexpensive. This may just be the Tanzanian form of *S. morsitans*. Adults tend to be lean and flattened no matter how much they are fed. It is sometimes confused with *S. polymorpha* because of the "tiger" common name (tons of *Scolopendra* have black striping and could be called "tigers"). However, this centipede is much smaller and it would take more than a dozen adult "African tigers" to make up the mass of any but the tiniest of adult *S. polymorpha*. It is sometimes sold as a "red headed" centipede even though two out of three have dark green heads. "Tiger" does not refer to behavior or voracity. They hardly attack anything other than appropriately small, teneral *Nauphoeta cinerea*. Adults sometimes live over a year, but these imports are short-lived and wimpy compared to almost any other centipede.

Scolopendra morsitans? African "Tiger" from Tanzania.

Scolopendra polymorpha Wood, 1861
Indications: Antenna composed of 21-27 segments, 7 or more smooth. First tergite with ring

Scolopendra polymorpha normally have seven or more smooth antennomeres. Here the first 13 are smooth.

Scolopendra polymorpha from Arizona.

Scolopendra polymorpha reddish form from California.

furrow. Tergite 21 with central groove. Length: 1.6-6.5″ (4-16.5 cm). Locality: Arizona, California, Idaho, Kansas, Montana, Nebraska, Nevada, New Mexico, Oklahoma, Oregon, South Dakota, Texas, Utah, Washington, Wyoming (USA), south into Mexico as far as Guerrero (Shelley 2002). Coloration: Cream to dark tan or orange body, often with an orange head and first tergite. Usually there is black margin at rear of tergites from the first or second to the 20th. White, tan, or bluish legs.

Scolopendra polymorpha has long been one of the easiest and cheapest centipedes to acquire for U.S. hobbyists and specimens even show up regularly on overseas lists. This species is extremely common in most areas of the southwest though mostly as immatures. It does not take much rock flipping in somewhat barren landscape to uncover a few. I have also found them running in the outer perimeter of lights during the evening, probably attracted to the availability of prey. This is the most northerly ranging *Scolopendra*, but it is rarely seen in the northern extent of its range and available specimens are nearly always from Arizona, California, or Texas. I have also seen specimens from New Mexico, but not yet any of the other states.

Scolopendra polymorpha laying eggs. © David T. Anderson

Scolopendra polymorpha from Arizona.

Captive hatches from wild-caught females are common, but the tiny young can experience high die off (over the first three to four months). Brood sizes are usually thirty to forty. The vast majority of immatures seen for sale have been collected rather than hatched. I observed a fully mature female from California lay eggs at just 1.6″ (4 cm) and a female from Texas at over 5″ (13 cm) but they were both fully grown adults. The small forms do not continue to grow in mature molts and the large forms do not reproduce when they are small

15.7 cm is the upper limits for size in *Scolopendra polymorpha*. Yellow form ("golden *heros*") from west Texas, USA. © David T. Anderson

Most *Scolopendra polymorpha* start out looking like this.

Scolopendra polymorpha California yellow unbanded.

Scolopendra polymorpha gold unbanded from southwest Texas, USA.

Scolopendra polymorpha large
blue adult "pseudoviridis."

Scolopendra polymorpha unbanded
blue form from California. © John Skierski

because they are not yet mature. *Scolopendra polymorpha* is not especially fast or aggressive and is slow to bite even when held (not grabbed). A nice specimen is larger than the many of the exotic "giant" centipedes. The body shape of immatures resembles other *Scolopendra*, but adults grow rather fat until the head seems too small for the body (the antennae and legs often seem undersized).

The only consistency in color patterns are the small immatures from various areas tend to have blue legs so the juveniles usually look a bit different from adults. However, in some areas the adults are just as blue as the brightest immatures. Whitish-blue adults have been sold as "Scolopendra viridis," "S. polymorpha viridis," and "S. polymorpha pseudoviridis," but all of these are incorrect. Some specimens from southern California seem to be very different from other populations since they grow large, are consistently unbanded, have dark colored heads, and the young are blue with abnormally long terminals. However these animals (often sold as "Scolopendra aztecorum")

Scolopendra polymorpha yellow
from California, USA. © John Skierski

morphologically do not even qualify as a subspecies since the adults fall well within standard deviation for this species. Unbanded yellow specimens, growing just as large, from southern Texas have been falsely sold as "Scolopendra heros," and there are banded specimens from Texas and Arizona that also have exceeded 6″ (15 cm). The massive difference in color patterns and sizes do not allow for geographic subspecies designations in any areas because there is so much variation. Even

Scolopendra polymorpha yellow from
California, USA. © David T. Anderson

Scolopendra polymorpha greenish with banded legs, Arizona, USA.

Scolopendra polymorpha with a nice red head and first tergite.

Scolopendra polymorpha from Texas, USA.

Scolopendra polymorpha "Aztecorum" collected between San Diego and Anzo Borrego, California, USA. © John Skierski

Scolopendra polymorpha "Aztecorum" immatures have unusually long terminal legs (not outside the normal range for this species) and are a brilliant light blue. Adults retain the dark head coloration but develop shorter, more standard, terminals and fade to yellow like most *S. polymorpha*. © John Skierski

within populations there so far has never been a single feature or even a definite stable pattern to stand on. Generally unbanded forms such as those found in California make up a small percentage of a larger population including mostly banded ones, but in some areas unbanded may be the common or only form. The unbanded specimens from California are usually light blue, but they can also be pale orange. Pale unbanded forms are not transitional colorations since the animals do not molt out of it.

Scolopendra subspinipes Leach, 1815
Scolopendra subspinipes subspinipes
Indications: Antenna composed of 18 or 19 segments, inner 6 smooth. No grooves on the head plate. Ring furrow absent. Leg pairs 20 and 21 without dorsal spurs. Terminal prefemur with two or three spines underneath as well as two or three facing inward. Length: 4.5-8″ (11-20 cm). Coloration: Brown to red-brown body, seldom with black margins. Dull orange-brown to red head and legs. Young animals (less than 4″ or <10 cm) can be black with orange head and legs or have an orange body with an incredible blue head and legs. Locality: Russia, India, Sri Lanka, Malaysia, Singapore, Laos, Vietnam, Thailand, China, Hong Kong, Japan, Philippines, Indonesia, Sao Tome, Principe, Ivory Coast, Liberia, Zanzibar, South Africa, Seychelles, Reunion, Mauritius, Rodrigues, Madagascar, Andaman Islands, Nicobar Islands, Bermuda, Central America, Colombia, Guyana, French Guyana, Surinam, Brazil, Caribbean Islands (including Puerto Rico and Dominica), and Pacific Islands including Hawaii (Kronmüller 2012). All the New World localities are considered introductions by man, while Hawaii certainly is.

Scolopendra subspinipes from China.

In general, a number of other species had been reduced to subspecies of *Scolopendra subspinipes* (*S. dehaani, S. japonica*, and *S. multidens*) and later returned to species level, though some have been discounted altogether. *Scolopendra s. subspinipes* is the most commonly available. Large brown orange-legged

Malaysian *Scolopendra subspinipes* pair. © Chia How

Scolopendra subspinipes from Malaysia molting. © Chia How

Scolopendra subspinipes, China. © John Skierski

Immature *Scolopendra subspinipes* in Maui, Hawaii, USA. © Peter Clausen

specimens arrived from China for years, some smaller, multicolored and banded specimens come in from Barbados now and then. There have been a number of brown to pink *S. subspinipes* in which the antennae and legs fade to a dark red toward the extremities. These imports have shown up as "Burmese," "Indo," "Malaysian," "Thai," etc., pink tips.

Large specimens are quick and powerful. If the hobbyist enjoys pets that seem to hate their owner, a huge *S. s. subspinipes* is the perfect choice. They smack forceps with their heavy body before they pinch with their rear legs and bite. When touched they rapidly curl the body into a C-shape with both fangs and terminals attacking in concert. They often sit very still in hopes of picking the perfect moment to burst into a flurry of biting anger. Many imports have damage (large black scars under the head or body) and six months to a year survival is average. Captive-reared adults can live up to four years after reaching maturity. Broods are rarely more than forty. Captive-hatched immatures can be easy to rear, but 20-40% die-off in the early instars is not uncommon.

Some forms start out at 2nd instar similar in color to the adult, while others have distinct juvenile color patterns. Young Hawaiian *S. subspinipes* are brightly colored with a blue

Scolopendra subspinipes, pale coloration. © John Skierski

Scolopendra subspinipes terminals. © John Skierski

Scolopendra subspinipes from Johor, southern peninsular Malaysia; also common in Singapore. © Chia How

Scolopendra subspinipes from southern Malaysia. © Chia How

Scolopendra subspinipes from Barbados.

Scolopendra subspinipes third instar.

Scolopendra subspinipes Malaysian "Red Tip." © John Skierski

head, first tergite, and legs, on a yellow-orange body (Clausen 2008), but the older immatures and adults are dark-reddish brown overall and look similar to the *Scolopendra subspinipes* pictured from Malaysia and Barbados.

Scolopendra subspinipes mutilans Koch, 1878
CHINESE REDHEAD
Indications: Antenna composed of 19 segments, first 6 smooth. Ring furrow absent.

Second instar brood and mother, *Scolopendra subspinipes mutilans*. © Kelly Swift

Terminal prefemur with one or two spines underneath as well as one facing inward. Length: 4-5.5″ (10-14 cm). Coloration: Black body, head and first segment red, yellow or orange-yellow legs. (Coloration very much resembles our *Scolopendra heros castaneiceps*, but the second tergite is not red and the terminals are reddish.) Locality: China and Japan.

Scolopendra subspinipes mutilans Orange Leg. © John Skeirski

Scolopendra subspinipes mutilans.

Scolopendra subspinipes mutilans.

Scolopendra subspinipes mutilans juvenile coloration. © Zhangjunduo

These behave very differently from any of the forms or subspecies kept so far and different from most any *Scolopendra*. They can be housed communally, practically on top of each other, and do not attack each other. They usually make no attempt to bite the keeper unless pinned. Furthermore, I have never seen one display the standard attack "C-shape" when touched with an object or finger. Wild-caught adults (as well as what appear to be large immatures) are often short-lived, with six months being good. A good number come in very lethargic because they are dying rather than because it is such a minimally aggressive creature.

Scolopendra viridicornis Newport, 1844
Indications: Antenna composed of 17 segments, first 4 smooth. Tergite 21 with narrow triangular ridge. Length: 6-8″ (15-20 cm). Coloration: Solid orange to black, with or without dark margins, and with or without banded legs.
Locality: South America

This was a beautiful and unique animal that was available in the mid-1990s but difficulties in identifying specimens have lead various people to suggest it never was here, since material from the old imports is no longer available. However, some of the specimens labeled *Scolopendra gigantea* "robusta" and *Scolopendra galapagoensis* were this species, while those labeled *S. viridicornis* were not. Fortunately, one of the defining characteristics (the flattened triangular ridge near the base of the last tergite) is often easy to pick out even in some old, lower quality photos. This is often a big, bulky species that can grow very fat. Also, like *S. gigantea* and *S. galapagoensis,* it may refuse to feed below 77° F, but unlike those species its babies do not eat their mother.

Scolopendra subspinipes multilans juvenile coloration. © Zhangjunduo

Scolopendra viridicornis. "This one I found in Suriname. It was funny, it was actually my guide who found it. He told me to come and have a look, he thought he had found a spider. Only the terminal legs were poking out from under the bark. I thought as soon as I lifted the bark, the centipede would run away, but it stayed there" (Martin Gamache pers. comm.). © Martin Gamache

Scolopendra viridis immature from southern Florida, USA.

Scolopendra viridis Say, 1821
Indications: Antenna composed of 21-27 segments, 6 or fewer smooth. First tergite with ring furrow. Length: 3″. Coloration: Emerald green body and yellow legs. Locality: Southern half of the USA northwest to Nevada, through Mexico and Central America (Shelley 2002).

Scolopendra viridis eating its exuvium. © John Skeirski

Scolopendra viridicornis, Peru.

Scolopendra viridis holding eggs. © John Skeirski

Scolopendra viridis is a pretty, shy centipede that likes to stay hidden. Body width is never over 1/4″ (6 mm). Immatures grow quickly and at all stages are excellent feeders. Adults are hardy but only live a year to a year and a half in captivity. Half-grown crickets are the largest food a mature animal can handle. This species is seldom offered for sale—hobbyists who buy animals labeled as *S. viridis* are sometimes buying *H. marginata* or *S. polymorpha*.

Akymnopellis chilensis (Gervais, 1847)
Indications: Antenna composed of 19 segments, first 5 smooth. Ring furrow absent. Legs without proximotarsal spurs. Length: 3-3.5″ (7.5-9 cm). Coloration: Dark greenish-blue body with brown marking towards head, legs cream fading to greenish blue at extremities. Locality: Chile.

Until recently the three *Akymnopellis* species from southern South America were placed with our North American *Hemiscolopendra* (Shelley 2008), found practically on the other side of the globe. They hardly resemble the North American native since they have no ring furrow and an unusual body shape, and they grow much larger. The body thins out decidedly towards the front in large specimens. These are remarkably calm for a centipede but become active on the surface each night.

Hemiscolopendra marginata (Say, 1821)
Indications: Antenna with 17 segments, all smooth (specimens Autauga County, Alabama), as few as 6 smooth by location/region. Ring furrow present on first tergite. Members of the genus *Hemiscolopendra* have no ventral proximotarsal spines on any legs, unlike *Scolopendra* (ventral spines are visible prior to the tarsi on large specimens but magnification is generally needed to see these spines on *Scolopendra* under 4″ (10cm)). Length: 1.7″-2.2″ (4-6 cm).

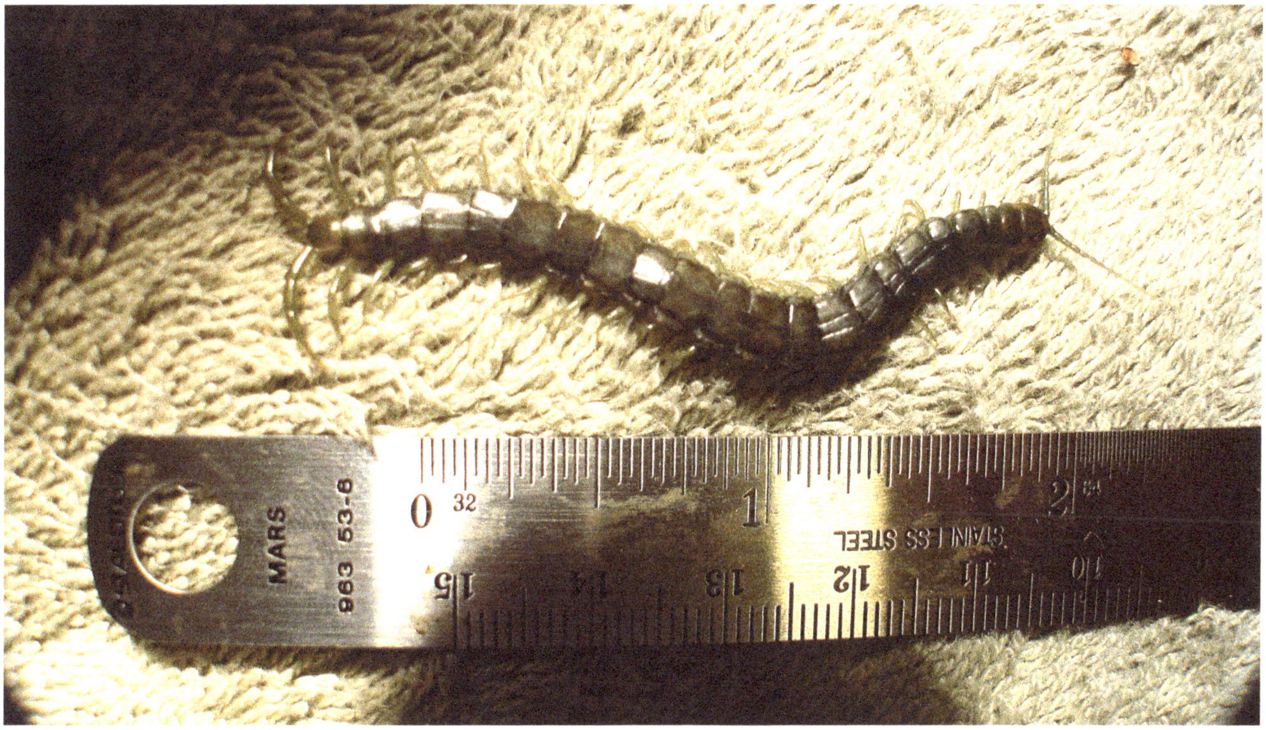

Akymnopellis chilensis, Chile. © John Skierski

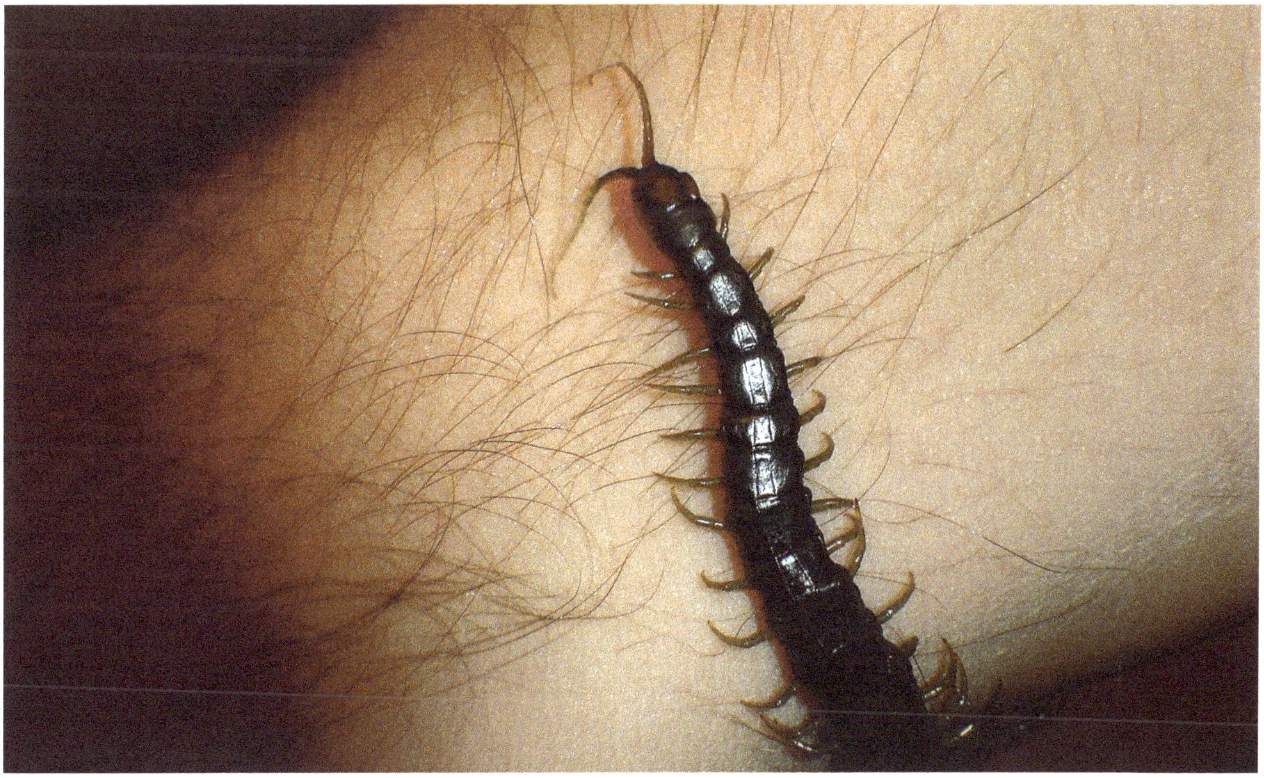

Akymnopellis chilensis. © John Skierski

Coloration: Light gray to darker blue-gray with dark tergal margins. Legs white with blue band near tip, last 3-4 pairs with blue bands on all leg segments except prefemora. Range: Much of the USA—Alabama, Arkansas, Arizona, Florida, Georgia, Illinois, Indiana, Louisiana, Kentucky, Mississippi, Missouri, Nebraska, North Carolina, Ohio, Oklahoma, Pennsylvania, South Carolina, Tennessee, Texas, Virginia, West Virginia (Shelley 2002, Mercurio 2010).

Hemiscolopendra marginata mother with first instars, Alabama, USA.

The species name refers to the dark colored margins of the tergites, whereas margins often refer to the raised lip on the edges of the tergites in centipede identifications. There is some variation in color by geography, moreso photography, but specimens can usually be identified by the pattern. *Hemiscolopendra marginata* is a nice little animal sometimes confused with the larger and prettier *Scolopendra viridis*. It has a much wider range and is far more common within its range than *S. viridis*. I have found specimens under bark of very old, rotten logs where there is half an inch of dirt beneath the bark. Swift (1997) reports finding a female with young (that were subsequently cannibalized) under the bark of a fallen tree.

This species is a good eater but specimens rarely, if ever, look fat. Gravid females are visibly bulkier than males but they hardly look rotund, even just before laying eggs. Although this species is rather easy to breed it is short lived as adults seldom survive past twelve months. Immatures take only a year to reach maturity at room temperature with adequate feeding. Keep in mind, the 2nd instars are tiny. They are small enough to climb nearly invisible films of dirt or moisture to escape through tiny gaps under the lids of some plastic containers. The first time I hatched out this species, nine days after the molt to second instar, there suddenly were only two survivors left from two dozen eggs. I had kept the mother in a 6-quart plastic shoebox (13″ x 8″ x 4″ high or 33 x 20 x 10 cm) and realized cannibalism was not the culprit when I found two second instars in the substrate of a different plastic shoebox on the same shelf a week later. I thought I had the problem solved the next time by keeping gravid females in 32 oz deli-cups (with a far smaller footprint of just 3.5″ or 9 cm diameter). The deli-cup had a tight fitting lid and just a few pinholes for ventilation. However, I then learned the second instars have to be moved out about day ten after the molt or they cannibalize rapidly (the mother might even eat some as well).

SUBFAMILY OTOSTIGMINAE

Indications: Spiracles rounded or ovoid, not valvular. Largest pair of forward spiracles are oval as usual for a member of the Otostigminae. (In Scolopendrinae the spiracles are narrow, slit-like, and resemble an isosceles triangle lying on its side and pointing toward the rear.)

Alipes grandidieri (Lucas, 1864)
FEATHERTAIL CENTIPEDE
Indications: The terminal legs are greatly elongated and end in a flat paddle with feathery markings. Seven longitudinal ridges—including the margins—run lengthwise down the tergites. Length: 3.5-4" (9-10 cm). Coloration: Chocolate brown body with yellow-orange legs, reddish terminals and antennae. Locality: East Africa, Tanzania, Zanzibar. The genus is common around the shores of the rift lakes.

The happy *Alipes* couple.

The most unique giant centipedes are without a doubt the noisy species in the genus *Alipes*. They look like normal Scolopendrids except the rear legs are highly developed and expanded into paddles with feather-like markings. The bizarre rear legs are used to make squeaky noises likely used as a warning to predators. Though they are not aggressive, they supposedly can inflict a painful bite. In captivity their first choice is to make noises when harassed. The male and female have the same level of "tail" development so it is not likely to be important to courtship. The tails are shaken outwards at a ninety-degree angle for stridulation, not rubbed together. Animals with only one rear leg make the identical sound, but with only about a third the volume (McMonigle 2006).

I have kept this species for years and have not been careful, often prodding specimens by hand during maintenance, and have never been bitten, but I never picked one up. A friend regularly handled his specimens on display: "When I kept these they were a favorite for live creature shows. They are one of the more predictable centipedes in their movements and easy to handle in front of a crowd" (Skierski pers. comm).

The tails are secondary to the molt, like the wings of insects. The "feathers" emerge from the molt shrunken and over the next half-hour are pumped up with blood. Wild-caught animals are often missing one or both of the rear legs. *Alipes* are bulky, moderately quick, and voracious. Captive females have produced as many as three successful broods following a single mating at five to six month intervals (McMonigle 2007). Only adults can tackle a large cricket.

In the wild *Alipes* are found along the shores of African lakes, including Lake Victoria. They are found under stones and are solitary. *Alipes* move with the receding shoreline that is the result of seasonal drops in the lake levels. The shore often recedes twenty feet during the dry season (Barney Tomberlin pers. comm. 2006).

Locations of *Alipes* according to Attems (1930):

Alipes appendiculatus—Zambia, Zaire, Uganda
Alipes calcipes—Angola, Congo, Zaire
Alipes crotalus—South Africa
Alipes grandidieri—East Africa, Tanzania, Zanzibar
Alipes multicostis—Cameroon, Nigeria, along the Gold-coast to Sierra Leone

The feathertails pictured here came from Tanzania through an importer in California. The centipedes were marked down 75%, so I bought a dozen. The lowered price may have been because the feathertails arrived missing one or both tails. The worst was missing eight legs and the best missing two. They were also only half-grown, which was not a bad thing. Immatures are more apt to molt more often and grow back the legs. Adult centipede molts take place at greater intervals and some species (including *A. grandidieri*) may not molt in adulthood.

Most looked almost normal in a single molt, but one animal took four molts to grow back both tails correctly. None died from their injuries though one was dried out and dead on arrival. (Its container had very obviously been overlooked as the paper towel was bone-dry.) Each underwent three to five molts that occurred in the first eighteen months and none in the following twelve months. At the end of two and a half years I had kept three. Then, the biggest one died. Since there had been no changes in care and it had not molted in a longer time than the others, the most likely conclusion was old age.

Reproduction Reports

The last two looked identical but none of the feathertails had displayed any sexual dimorphism. I decided to put the two together; if they were approaching the end of their lives they deserved a shot at reproduction. More aggressive centipede species normally do not attack each other right away if given space, media to dig in, food, and are not just thrown in on top of each other. However, the risk of one being eaten increases with every passing night.

Alipes adults are under 4″ (<10 cm) trunk length so a 5-gallon aquarium was chosen. Setup included 1/2″ (12 mm) of aquarium gravel covered with 2″ (5 cm) of ground coconut shell. The substrate was kept very damp since even adult feathertails are vulnerable to desiccation. The lid consisted of a piece of glass with a 1 1/2″ square opening at one corner and a screen lid over top. Decorations were a section of cork bark and a length of driftwood. They crawled away to opposite corners after touching. I checked on them a few days later and found they had carved out a shallow under the driftwood where they were resting wrapped together! Other species tend to have separate burrows dug in opposite corners after the first day or two. They were given four or five 1/2″ (12 mm) crickets twice a week and pillbugs were placed in the tank for clean up (some were eaten).

Tanzanian *Alipes grandidieri* with eggs.

They lived together four months before she ate him. It is not possible to prove she killed him. He may have died of old age though he had not shown any signs. (Commonly centipedes look a little dull or sluggish and lose their appetite weeks or months before death.) Whether he died from love, age, or something else, it was certain she ate him. Though often considered communal, cannibalism unrelated to hunger is not extremely rare for this species.

Less than two weeks later she was holding eggs! Approximately forty days later when the eggs developed into protonymphs, it appeared she had eaten a successful mating partner. Protonymphs are a lot bigger than the eggs and in this genus they resemble a tire. (*Scolopendra*, *Hemiscolopendra*, and others tend to look like croissants at this stage.) Progress was checked every week in order to take photographs and gather data for a developmental timetable. Each check was a relief because she had not eaten the immatures, but worrisome since each time could be the disturbance to upset her. Each time the female was carefully nudged back into position under the driftwood and substrate placed over the sides of the hollow to seal out light and intruders and keep in humidity. Following thirty days, the protonymphs hatched into first instars, which look like normal centipedes except they are yellowish white and do not eat. The female holds onto the first instars until they molt, as is common for Scolopendromorpha and some Geophilomorpha. (In other centipedes the first instar is solitary, fully mobile, and may be able to feed.)

Less than twenty days after the change to first instar, tiny antennae were waving from below the dirt at the front of the cage. They had finally molted and had quickly abandoned the mother; they had been first instar two days earlier. Total time from eggs to second instar was approximately eighty days. Two thorough searches uncovered a total of thirty-two second instar *Alipes*. Each was moved to a 5.5 oz cup with one thumbtack hole in the lid and 1/4″ (6 mm) of wet coconut fiber. Three young pillbugs about 2 mm and six to ten fruit flies, *D. melanogaster*, were placed in each cup. By the next day most had eaten every fruit fly. It was amazing how many fruit flies they could go through and most ate more than fifty in the first week. Two died on day seven and another pinhole was added to the lids in case the ventilation wasn't adequate for the constant feeding.

Alipes grandidieri terminals in use. This specimen molted a few days earlier and regenerated the white-colored leg below the right side of the seventeenth tergite.

On day twenty-three the first one molted to third instar. *Alipes* at this stage are close to 2″ (5 cm) including the tail, which is bigger than L3 for some of the biggest centipedes. At this point it seemed they were too big for fruit flies and their food was switched to small lobster roaches, *Nauphoeta cinerea*, and firefly roaches, *Schultesia lampyridiformis*. Two to three roaches were added each week but only fireflies after the first few feedings since the lobsters weren't being eaten well.

It took two months for a few to finally molt to fourth instar. A third looked sickly and thin and the skinniest one died. The runts were offered some fruit flies and it seemed apparent

Alipes cohabitation usually does not end well. This particular scene may not take place for weeks or months.

Blue Ring mother with eggs. © Amy and Gabriel Tendera

the fruit flies should not have been stopped before they would no longer accept them. One or two lobster roaches that still had not been eaten were removed from half the cages. In two weeks every single one molted to fourth instar. Previously, it had seemed the fruit flies annoyed the little centipedes because the feathertails went running around the cage swinging their tails wildly whenever a group of fruit flies was dumped in. After seeing how helpful resuming the flies was, the reaction translated differently, like puppies wagging their tails in excitement at feeding time. (C. Schwarz warns the reader to disregard this anthropomorphism.) The first two molted to fifth instar after another three weeks (seven weeks since the first molted to fourth instar).

The female began to eat as soon as the young were gone but after a few months her appetite decreased markedly and it seemed she could be constructing a new hollow under the driftwood. Was she getting ready to molt, lay a new clutch, or getting ready to die? Seven months ten days after the first eggs were laid, she was holding a new batch! She had not molted in two years (it is possible this species does not undergo postultimate molts). Most centipedes continue to molt after maturity, but *Alipes* might not be one of them. Forty days later the second brood changed into protonymphs, but this was not the final chapter. Five months later, the same female was again holding a batch of fertile eggs. The third brood was a little larger with thirty-six second instars. Unfortunately, a few months after this last brood dispersed, the female died.

Ethmostigmus trigonopodus

Blue Ring Centipede

Indications: Antenna usually composed of 19 or 20 segments, 4 smooth. Length: 4-5.5" (10-14 cm). Coloration: (Tanzania) army green to brownish green body, legs usually cream with light blue bands on each segment. For the first few molts, young are emerald green overall. The blue rings on the legs begin to develop at fourth instar. Nigerian specimens are nearly black with solid orange legs (Lewis 1981), there are likely many color forms across its range. Locality: Malawi, Nigeria, Tanzania, Zambia, and much of Africa.

Ethmostigmus trigonopodus, 2002, Blue Ring Centipede.

Specimens from Tanzania are very common in captivity and are best known as the blue ring centipede in reference to the blue bands (rings) on the legs. Sometimes the rings are bright blue, but more often green or bluish-green. Specimens are calm and slow but they do like to grab onto and test everything with the maxillipeds. Most of the time specimens sit on top of loose substrate, but in deeper, compactable media they completely bury themselves for months. Captive hatchings are very common and the survival rate of young after dispersal with moderate care is nearly perfect. Even the wild-caught adults are generally long-lived and four to five years in captivity is not unusual for an animal that was already mature. This is a good beginner's species but it is not terribly common nor as inexpensive as it once was.

After rearing a group individually on shallow substrate in cups, I placed four 3.5" (8 cm), captive-hatched, two-year-old immatures together in a 10-gallon (40 L) terrarium with 6"

Ethmostigmus trigonopodus, Blue Ring Centipede. © John Skierski

Ethmostigmus sp., "Giant Yellowleg."

An *Ethmostigmus* sp., "Giant Yellowleg," should never be skinny except shortly after a molt.

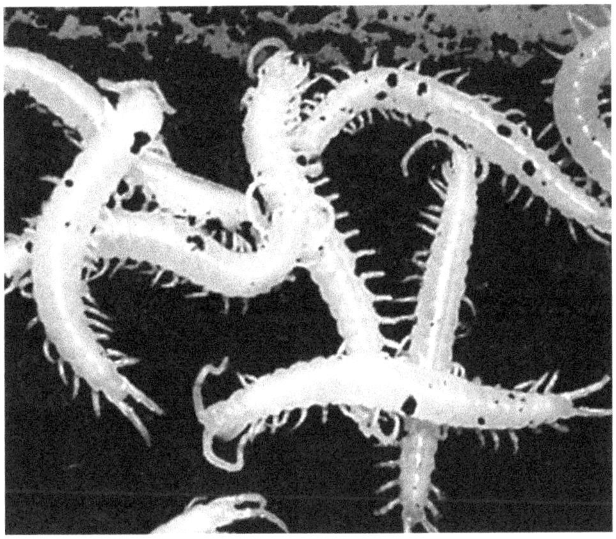

Ethmostigmus trigonopodus Blue Ring first instars taken from cannibalizing mother.

(15 cm) of damp, sticky compost. The substrate was very damp, somewhat pasty, and if a finger was pushed in and removed, the hole would remain. The blue rings were fed mostly crickets along with some *Nauphoeta*, *Blaberus*, and isopods. The cockroaches were fed sparingly since these burrowing species could hide in the substrate and provide a possible hazard if a large number were uneaten. The centipedes lived together for four years, all reaching maximum size for this species within the first year. Unfortunately I almost never saw them. Every three months or so I would dig up the tank to see if any were still alive.

About two and a half years in, when digging through, I discovered a female holding a large group of 1st instars. She was inside a large cavern about the size of my fist. I was incredibly happy because these were my first true captive-bred specimens (born from captive-hatched and reared, rather than wild, parents). The female *E. trigonopodus* did not share my happiness at all. She began to consume her offspring, one by one, in front of my eyes. I had originally tried to prop up the walls and fix the chamber but her brood chamber was essentially destroyed. After the third or fourth larva was drawn into her mandibles like a spaghetti noodle, I took the thirty-two remaining first instars from the mother. I was certain she would have eaten the entire group over the next fifteen minutes so there was nothing to loose if they could not survive without her. The first instars were placed on a thin layer of damp soil from the cage. Things looked promising as they seemed fine for a week and then one dropped over dead, the next day another, and so on. About ten survived to the first molt, three and a half weeks in, but seven of the survivors did not molt properly and died halfway out of the old exoskeleton. Two seemed to molt fine and then fell over dead. This was not the result of lack of moisture, as they were kept damp. The single second instar to molt perfectly and survive was a little sickly for the first few molts but eventually grew into a healthy adult after just over three years.

Ethmostigmus sp.

Tanzanian "Giant Yellowleg"

Indications: Antenna with 19 segments, first 4 smooth. The coxopleuron is longer and with reduced spines compared to above. Length: 4-6″ (10-15 cm). Coloration: Greenish body, yellow legs and antennae. Locality: Tanzania, Africa.

This species is similar to the blue ring centipede and the two are often imported together. However, whether they come from the same or different areas in the collection range is uncertain. Many captive-produced young have been reared and none have developed blue rings on the legs (or vice versa). The bite of this species is supposedly far more potent and painful and specimens appear more defensive and likely to bite. Adults are long-lived but no stages are quite as hardy as the blue ring centipede. Lastly, the adult males do not grow incredibly heavy and generally do not look like the fat-bodied adult female.

Blue-legged Otostigminae from Peru.
© Frank Foldbjerg Hansen

Again, like the blue ring centipede, adults can be kept safely together without cannibalism. If placed in a small cage with a shallow layer of substrate, safe cohabitation is probably unlikely. However, I maintained three large, wild-caught adults in a 5-gallon (20 L) bucket with 4″ (10 cm) of damp, compactable substrate for four years (only a gallon or so of total substrate for all three adult centipedes). They were full-grown and grew fatter, but the armor and legs never grew larger. Unfortunately no signs of reproduction were seen though they were dug up inside of large, oval chambers on a few occasions.

Otostigmus scaber Porat, 1876
Indications: Antenna with 20-22 segments, first 2.5 smooth. Tergites 3-20 with distinct longitudinal ridges. Length: 3-4″ (7.5-10 cm). Coloration: Very striking with alternating blue and white legs and terminals. Locality: Asia including China and Japan, also Hawaii, USA (adventive).

These are good eaters that take out nearly any type of smaller prey. No matter how much they are fed they remain thin-bodied like *Alipes*. They are rapid burrowers that utilize deeper substrate

Otostigmus politus yellow form from Beijing, China. This species grows to plus or minus 3″ (7-8 cm). © Zhangjunduo

Otostigmus scaber from Jiangsu, China. © Zhangjunduo

forming a network of tunnels. The beautiful terminals are held in the air like warning flags when the animal is threatened. Females tend to be larger and do not have the notable, inner, finger-like, terminal prefemoral projections. This species is a common adventive in Hawaii and came in from a Japanese export at least once. Presently, Hawaiian Fish and Game has no qualms about export of non-native, invasive centipedes from the islands, but nobody offers this species commercially. All native (protected) Hawaiian centipedes are in the order Lithobiomorpha, but of course the most commonly seen stone centipedes are adventives as well.

Parotostigmus rex Chamberlin, 1914
Indications: Antenna with 17-18 segments, first 3 smooth. Tergites with faint longitudinal ridges. Length: 3.5-4″ (9-10 cm). Coloration:

Otostigmus scaber from Hawaii, USA. © John Skierski

Otostigminae, Peru. © Frank Foldbjerg Hansen

Parotostgimus rex from Peru. © John Skierski

Parotostigmus rex terminals. Females like this have no accessory appendages on the inner prefemora. © John Skierski

Black body with red head and first and last tergite. Legs yellow except for red terminals. Locality: Peru

This handsome species from Peru was imported in 2008 and females produced offspring successfully, but further generations have not been reported. The young proved to be very easy to keep and rear with excellent survival, but I traded mine away when they grew to 3″ (7.5 cm) at just under two years of age. This species is reminiscent of *Alipes* in body shape, though the color pattern resembles its red-headed cousins from North America, *S. heros castaneiceps*, and Asia, *S. subspinipes mutilans*. The juveniles are colored differently from the adults, overall pale brown like dirt, but they start to approximate the adult pattern after a few molts. Males and females are very easy to sex because the males have a large, finger-like projection on the inside of each terminal prefemur. As is common for many Otostigminae, adults seem relatively tolerant of each other though cannibalism has occurred.

Rhysida longipes from Shenzhen, China. © Zhangjunduo

Rhysida longipes (Newport, 1845)
Longtail Centipede
Indications: Antenna with 18 segments, first 3-4 smooth. Short coxopleural process with two short spines. Terminals long and thin. Terminal prefemur with three inner and three lower spines usually. *Rhysida* has spiracles on segment 7, while *Otostigmus* does not. Length: 3-3.5″ (7.5-9 cm) Locality: Africa, Asia, Mexico, Florida, USA (Introduced).

Rhysida longipes is sometimes listed as the African olive or African longtail centipede, but specimens seen in captivity are normally collected in the United States, in southern peninsular Florida. This adventive species was recorded in Florida by Wood (1862), however it is a widespread species in Africa and Asia, so it must have been introduced by humans sometime before 1862. The brood size for one Floridian specimen was 42 eggs. Wild-collected specimens from Florida are readily available,

Otostigmus politus Blue Leg from Beijing, China. © Zhangjunduo

Otostigmus aculeatus from Shenzhen, China. © Zhangjunduo

Rhysida longipes adult female from Florida, USA.

Rhysida longipes laying eggs.

but there is little demand. It is a rather handsome, heavy-bodied creature for its size, including mottled green tergites and mottled legs and spectacularly elongate terminals. Specimens are hardy and live multiple years in captivity. *Rhysida* is the second largest genus of the Otostigminae, with around forty species in the New World tropics, the Indo-Malayan region, and much of Africa (Bonato et al. 2011).

Scolopocryptops spinicaudus from coastal Washington, USA. "These could be virtually anywhere in western Washingon. I know for sure I have collected in Pierce, King, Kitsap, Pacific for WA, excepting the greenish specimens from Klamath County, OR. The largest I`ve seen was from Thurston County" (Skierski pers. comm.). © John Skierski

FAMILY SCOLOPOCRYPTOPIDAE
Indications: Twenty-three leg pairs, no eyes.

SUBFAMILY SCOLOPOCRYPTOPINAE
Indications: Terminal legs are straight and thin with a large inner spine on the prefemur.

Scolopocryptops spinicaudus
Coloration: Orange to red-brown. Length 1.2-1.6" (3-4 cm). Locality: Western coast of the US from Alaska, through Canada, to southern California, Japan (Shelley 2002).

Half a dozen *Scolopocrytops* are found across much of the U.S., though this and *S. sexpinosus* are the only species ranging into Canada. They are all similar in size and appearance, though there are some Asian species in the genus which are a little more colorful. They are similar in care to the larger species.

Scolopocryptops spinicaudus, a smaller green-tinged species from the Oregon–California border between I-5 and Klamath Falls, USA. © John Skierski

ORDER SCUTIGEROMORPHA

Scutigera coleoptrata molted off surface.

Indications: Large compound eyes, all thirty legs at least four times longer than body width.

The house centipedes are divided into three families: the South African and Madagascan Scutigerinidae, and the Neotropical–Afrotropical Pselliodidae and Scutigeridae (Edgecombe and Giribet 2006). Members of the different families look quite similar to the observer but there are differences in the segmentation and sometimes maxilliped shape that can be seen just barely with the naked eye. Also, the number of eye facets can be different by species. Each eye of a mature *Scutigera coleoptrata* contains more than 200 ommatidia, while some, such as the Philippine species shown here, possess more than 500 facets in a single eye—the species *Scutigera araneoidea* Grenacher, 1880 according to the description has just 100. The three small species in the Family Scutigerinidae have less complex eyes and reduced appendage segmentation, while the differences between the Pselliodidae and Scutigeridae can require molecular analyses. There are only three species of Pselliodidae with the vast majority of species in the Scutigeridae.

Representatives of the order Scutigeromorpha are oddities. They are unable to burrow like other centipedes and have an adapted respiratory system with spiracles on the top. (Spiracles are located on the sides of the body in all other centipede orders.) The blood aids in transporting oxygen. Some tergites are reduced and others expanded so much that there appears to be just eight from above versus the fifteen leg-bearing sternites visible from beneath. The overlapping tergite structure reduces wave movements in the body which allows for faster running but also eliminates the flexibility of the body so important to the movement of other centipedes. House centipedes are impressive creatures that

Handling a harmless house centipede.

appear to be made out of legs. The legs are far longer than anything seen in the other orders and are outfitted with long spines and hypersegmented tarsi. The component tarsomeres are so numerous and small they are difficult or impossible to see with the naked eye. The antennae are even more extensively segmented with as many as 400 antennomeres. The eyes look like "bug eyes" since they are made of numerous ommatidia and resemble the compound eyes of insects and certain crustaceans. Even the behavior differs from other chilopods. They can see and react to the approach of predators, and the calculated prey attacks contrast the blind thrashing movements of standard centipedes.

Although they are unable to capture large prey like other orders which wrap the flexible body around the prey while holding on tightly with the legs, house centipedes have their own methods. They would be restricted to tiny and select prey if the long flexible maxillipeds and palps were the entire prey-catching mechanism. The hypersegmented tarsi are used like living ropes to encircle small prey appendages. They wrap around the antennae and legs of prey in an unbreakable loop. Without magnification it is difficult to figure out how they are capable of hanging onto large prey.

Scutigeromorpha centipedes are masters of vertical surfaces. In the wild they have been found living on tree trunks, rock faces, and cave walls. In homes they are usually encountered running up and down the walls rather than across the floor. If chased they can jump from the wall and run across the floor at top speed. Young centipedes more readily parachute than adults. Due to the minimal weight and drag provided by the outstretched legs, they almost seem to float as they slowly descend to the floor.

Scutigerid from the Philippines.
© Martin Gamache

Scutigerid from the Philippines.
© Martin Gamache

Malaysian scutigerid. © Chia How

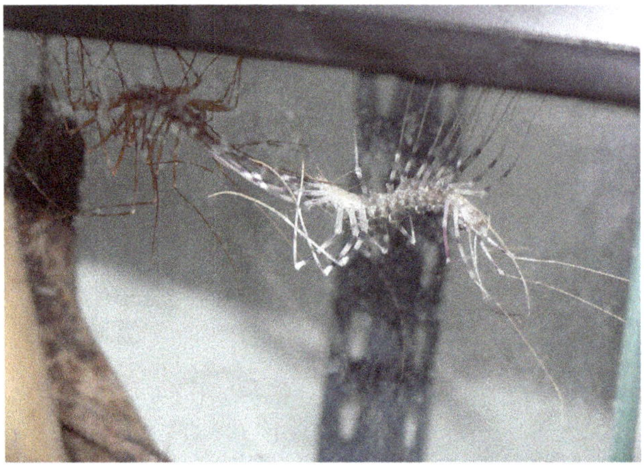

Scutigerid molting while hanging from the lid, larger Malaysian species. © Michael Fanto

Although somewhat tolerant of conspecifics they are generally solitary and prone to cannibalize over time. While it is not rare to find two or three half grown immatures under the same small rock, they eat each other if confined, especially small immatures. Adults rarely may be found under cement bridges or cave entryways in small groups, but they are seldom close enough to touch antennae and would eventually cannibalize if confined in a cage.

Possibly the most interesting aspect of this order is that members comprise the only invertebrates that can molt off a flat substrate or hang from surfaces during the molt. With the only exceptions based in semantics, all the other invertebrates either must molt while hanging or on top of a surface depending on species or order. For example, all whipspiders (Arachnida: Order Amblypygi) must hang while molting and all whipscorpions (Arachnids: Order Thelyphonida) molt on the surface (inside dug out chambers). Molt orientation can be consistent within an order, although spiders can molt while hanging from or on a surface, but not the same species. Even the spider groups that molt off the surface, such as tarantulas, still lay on their backs suggesting they once hung upside-down from webs. Scutigeromorpha not only molt from upright or upside-down horizontal surfaces, but also any angle up to and including vertical.

Autotomy is highly advanced. This is the only group of centipedes whose legs seem to fall off at the barest of touches. Afterwards, the abandoned legs writhe and twitch for a few minutes. Where autotomy differs most significantly from other centipedes is the speed of regrowth. Shed legs are often regenerated quickly since they molt more often than other centipedes, but also more readily following damage. Other centipedes and arthropods such as phasmids and whipspiders take the same time or longer than siblings from the same brood to reach the next molt if legs have been autotomized (McMonigle and Clausen 2002, McMonigle 2013). Damaged legs actually speed up the process in Scutigeromorpha, assuming they are provided adequate prey. On multiple occasions when I have caught a group of half-grown *S. coleoptrata,* specimens that lost legs have molted within two weeks to regenerate the missing legs while none of the undamaged animals have molted in the same time frame. Since they do not eat the exuvium, molts are very easy to track. This highly refined defensive mechanism can also play into husbandry. When specimens are of ill health, mostly from being inadequately or improperly fed, they can begin to drop legs. If caught right away it is possible to improve feeding and save the animal. If, however, it has stopped feeding and lost multiple legs, there is no hope. Sick and poorly nourished animals do not always autotomize, they may just die in place and turn dark purple.

The incredible speed of house centipedes is likely employed for escape. Generally they do not chase down prey until it exhausts like a cheetah might, but rather visually track and pounce on prey as it approaches. Like many predators, they are attracted to the feast brought in by artificial lights in the evening. I have seen them running around catching prey at lights, but they do not seem to zero in on distant prey. They do seem capable of seeing me from many feet away, but the first response is often freezing in place.

House centipedes are restricted primarily to the tropics, especially of the Old World, though there are a few uncommon, small ostensibly native species, whose taxonomic validities have not been confirmed, found in the American desert southwest. This group would be entirely unfamiliar to residents of North America and northern Europe except for a single species, *Scutigera coleoptrata*. Most people are familiar with the common house centipede. (The Order Scutigeromorpha has the most standardized common name of all five orders because it is named for this ubiquitous species.) Adult *S. coleoptrata* seldom have a trunk length much over an inch, but the leg structure makes them seem gigantic. The truly large tropical species appear even more impressive. At least one such gigantic tropical species from Andhra Pradesh in southeastern Indian species can reach a trunk length of nearly four inches (~10 cm) (McMonigle 2003). A handful of specimens of a giant 2-3" (5-7.5 cm) species from Malaysia showed up on some vendor lists in the U.S. in 2009. Despite success hatching the eggs and rearing immatures through a few instars past the 30-leg stage (to approximately 10th instar), it seems none made it to maturity. More recently just two specimens of a ~2" (~5 cm) beauty from the Philippines showed up through a Canadian vendor, but scutigerids are rare imports.

Special Handling Considerations

Although cage cleaning is generally unnecessary, the scutigerids may still need to be rehoused as they grow, or for mating purposes. Smaller species like *S. coleoptrata* and *Dendrothereua* spp. are realistically unable to bite, but direct handling should be avoided because the animals readily throw legs by the dozen and a surprisingly light blow to the body results in rapid, convulsive death. To avoid injury, place the rim of the containers inside each other and gently coax the animal to walk over.

Nearly any physical contact results in missing legs. Forceps should never be used. Fortunately, adults molt every six weeks and legs grow back easily. Immatures can grow back legs even faster. A frighteningly maimed animal can look perfect in less than two weeks. Usually all the legs grow back at once, but this is not always the case. One *S. coleoptrata* grew back one or two legs each molt but then lost a different leg each time within a day or two of the molt. Finally, after half a dozen molts it had all its legs without dropping any. It later lost one of the rear legs during setup for a photo.

Better than House Spiders

For most people, the common house centipede is #1 on the list of inverts found in the home that sends chills running down their spine. Yet, it is not capable of delivering a painful bite to humans and like house spiders it is beneficial because it eats insects in the home. Prey includes mosquitoes and house flies as well as stored product pests like adult meal moths and carpet beetle larvae. Unlike spiders, house centipedes never leave messy cobwebs or paint the floor below with sticky, black and brown frass droplets. They also do their job while trying to keep out of sight, unlike various house spiders. Although it is rare to find popular literature extolling the virtues for the house centipede, Waldbauer (1998) explains: "If you see one of these creatures in your home try to keep yourself from squashing it. House centipedes do no harm, but are really beneficial because they eat cockroaches and other less welcome bugs that invade our homes."

"It (*Scutigera coleoptrata*) is harmless to humans" (Breene 1997). It does not bite you while you are sleeping like some house spiders (though a good number of so-called "house spider bites" are actually flea bites or bacterial infections; even bacterial hoof rot in horses was long blamed on spider bites (Baxter 1993)) and these centipedes normally cannot bite when handled with bare hands. Though a bite is not 100% impossible, factual reports are exceptionally rare and entail pain lasting one or two minutes at most. I have bare-handled many hundreds of specimens over more than a decade without a bite. I recently tried to force a bite from a large adult female by manhandling her and carefully pushing her venomous maxillipeds into the soft skin of my forearm near the inside of my elbow (to see if rough handling and thinner skin could result in a bite). The only pain she caused was deep regret for her rapid and nearly complete dismemberment.

Female genitalia close-up.

The last tergite almost entirely conceals the male genitalia. Freshly molted mature male.

GENDER AND MATING

Gender determination appears to be the same for all species. The sexes are obvious but I was unaware of the difference till I had both a mature male and female, and the female laid eggs. Size is not an indicator as the largest and smallest adult specimens I have seen of *Scutigera coleoptrata* were both males. From above, the terminal tergite (last segment) points downward on the adult male. The female's terminal segment is not as sloped and is followed by a small but notable ovipositor. It is possible to easily determine gender when immatures reach two-thirds full size, but not before then, because the female's ovipositor is not yet visible. All younger specimens appear "male." In captivity the only way to determine adulthood is when subsequent molts result in no growth (the female's ovipositor is visible, though smaller, a number of molts before maturity). Wild specimen maturity is easiest to determine by time of year as a result of seasonal reproduction. Sexually mature specimens continue to molt in captivity but do not continue to grow.

An initial attempted mating of *S. coleoptrata* took place in a ten-gallon (40 liter) terrarium with some wood and logs for decoration and hiding. Two adults were placed in the cage but I did not know how to determine gender at the time. They recoiled on brief encounters and seemed to show no aggression but on

Female genitalia.

Scutigera coleoptrata egg compared to the size of the female.

Male *Scutigera coleoptrata.*

This subadult male *Scutigera coleoptrata* might be mistaken for a female but the genitalia is relatively concealed by the last tergite.

The first instar has just eight legs.

The spermatophore is many times larger than an egg and looks similar to a *Scolopendra* spermatophore, but there is no webbing.

day sixteen one was missing in action. Later breeding attempts with sexed adults took place in 16 oz deli cups and the pairs were left to spend just one evening together or removed shortly after observed mating. Larger tropical species would need proportionately larger enclosures for mating. Upon introduction, the male usually backs to the female and tries to guide and direct her using his terminal legs. The spermatophore is large (~3 mm) and seems to be accompanied by little, if any, webbing.

OVA AND EARLY DEVELOPMENT

Females may only produce eggs twice in a lifetime as a result of short lifespan and seasonal oviposition. The tiny white spheres are approximately 1 mm in diameter and are deposited one at a time. Some seem to be glued into the dirt, but not thoroughly covered, while most are laid freely. (Lithobiomorphs place eggs inside clumps of dirt.) Eggs can be deposited on damp paper towel, but a 1/4″ (6 mm) layer of soil is suggested for egg laying. Only two eggs were ever laid on paper towel. It is possible the female laid many more eggs and consumed them, but those laid exposed on soil are never eaten. Eggs can be handled by human fingers without sustaining damage since they have thick, leathery shells compared to epimorphs like *Scolopendra*. The female should be removed from the cage when eggs are found, since she can eat the hatchlings. Eggs hatch into a mobile immature in approximately one month. The eggs do not enter a resting stage. The hatch rate for recovered eggs has been close to 100% unless allowed to dry. I have observed a number of hatchings and the Scutigeromorpha certainly lack a protonymph stage. The 1st instar stage does not eat but is fully mobile.

The large Malaysian species pictured produced nearly four dozen eggs over a one-month period, which were deposited in the substrate (Michael Fanto pers. comm.). I have produced

Young *Scutigera coleoptrata* are usually bright purple in any light.

Second instar *Scutigera coleoptrata*, compared to a hatchling cricket.

six broods from different *S. coleoptrata* females over five years and they have all laid two to three dozen eggs over the period of two to four weeks (except for the seventh female that oviposited only two eggs on paper towel). *Dendrothereua homa* seems to lay even fewer eggs, but captive-rearing data is very limited. Even indoors with constant feeding, egg laying seems to only take place once a year. In *S. coleoptrata* this occurs only from late spring to early summer, but tropical species may lay eggs at different times of year.

The first mobile immature has only eight legs and looks very much like a tiny cricket. The eight-legged creature does not eat and molts after about ten days into a ten-legged creature as it slowly begins to resemble the leggy adult. At this time the immature begins to feed. At each subsequent molt the animal gains four legs (two pair). The leg count on *one* side over the course of the instars is 4, 5, 7, 9, 11, 13, and 15 (McMonigle 2008). At 75° F, immature *S. coleoptrata* took ten weeks to reach 15. The young centipedes are a handsome light purple from leg counts 7 to 13, but they fade to blue and eventually tan. These instars also have fewer tergites, an unusually large rear, and no terminal legs. Upon reaching 15 (thirty legs) the immature is only about a quarter grown, but it does not continue to add legs with more molts. It will take at least another five molts to reach full size.

Molts slow down to once every three to four weeks, but can take much longer if fed minimally or if fed primarily deficient prey such as *Drosophila*. Molting frequency is incredible compared to other centipedes and even adults molt every six weeks when fed appropriately. The molt takes place on top of the media since climbing areas are not needed. The molt is not prone to difficulties. They remove themselves head first like other centipedes, but unlike other centipedes the old exoskeleton is not eaten, and is left to rot away.

Scutigeromorph enclosures
with large immatures and adults.

REARING

Members of the Scutigeromorpha feed on small living invertebrates and appear to be the only centipedes that are obligatory carnivores. Other centipedes can feed on vegetable matter and consume their own shed exoskeletons. *Scutigera coleoptrata* are capable of taking down prey half their size, but prefer prey only a few times the size of their head capsule. Scavenging on freshly killed prey is possible, especially if the centipedes are starving, but even in tight quarters dead prey usually remains untouched. Large prey might damage them but are generally just ignored. For the first few instars *S. coleoptrata* are too small to eat hatchling crickets or firefly roaches (*Schultesia lampyridiformis*). A captive culture of "giant" springtails (adults of this species are barely the size of very small *Drosophila melanogaster* fruit flies) has proven a useful food source since most other commonly cultured springtails are too small and burrow. Burrowing prey is almost never taken by scutigerids even if it surfaces at just the right

The last three molts are visible on the bottom of the enclosure. They are thin and eventually disintegrate.

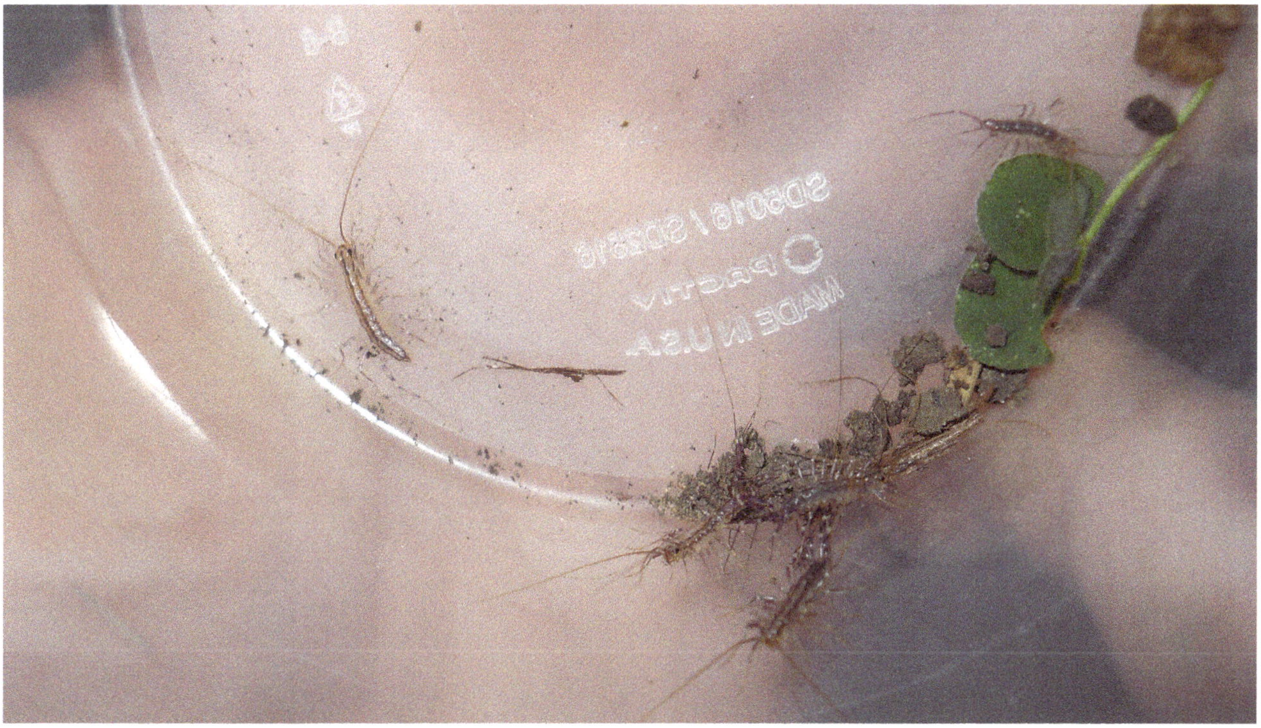

Wild-caught specimens often lose some legs. The two seen here, which autotomized a number of legs, molted after ten days and regenerated all missing legs.

This *Scutigera coleoptrata* is feeding on a *Nauphoeta cinerea* nymph.

Scutigeromorpha are highly segmented creatures.

Up close it is possible to see one tarsus wrapped around the antenna and another around the rear left leg of the cockroach.

time. Early instars are too small to prey on small fruit flies, but fruit flies should not be offered during early development in any case.

In *S. coloeptrata*, all unexplainable deaths have occurred in the first six instars while being fed on springtails, but it is impossible to know if the survival rate is limited for early instars, by captive husbandry technique, or as a result of prey item's lack of nutrition. Other cultured prey items are simply too big at first. After a few molts they can eat fruit flies, but when *Drosophila* are used as a singular food in the early instars the results have been 100% mortality. Survival between 7th instar and maturity has been nearly perfect with moderate care with exception of the Malaysian species (possibly the result of too much reliance on fruit flies later). The survival rate from hatching to maturity of my first *S. coleoptrata* brood was eight out of twenty-three separated (six were cannibalized prior). Later groups had 65-95% survival past separation, but fruit flies were used sparingly, if at all. Better understanding of feeding or closer attention may have been the cause of improved survival. Fruit flies are excitedly eaten and can be used as temporary food or partial diet for large immatures and adults. However, they lack important nutrients and the animals will not molt again until fed acceptable food. Fruit flies are like dirt cakes since they can delay starvation but do not allow growth. Adults can survive a long time on deficient food without molting but immatures fed only *Drosophila* flies will not live long. Second instars of large-bodied tropical species can be large enough to feed on hatchling crickets or newborn firefly roaches, which could lead to a higher initial survival rate.

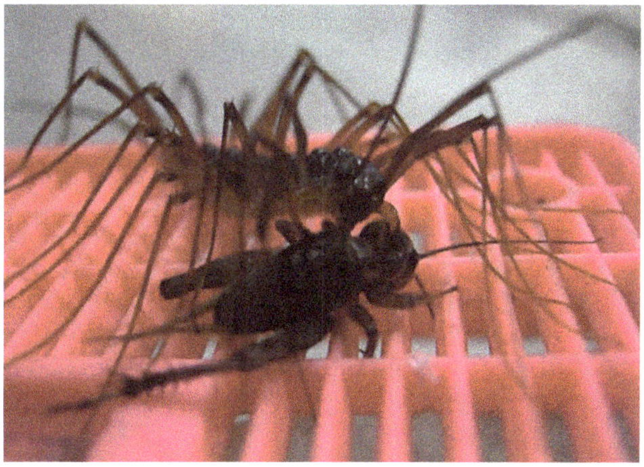

Smaller Malaysian Scutigeromorph feeding on a cricket. © Chia How

Communal housing is generally a terrible idea if the survival of more than a single specimen is desired. Hatchlings can be kept together for the first three or four instars but then they begin to cannibalize at an alarming rate despite available food and space. *Scutigera* begin to cannibalize at 3rd instar but intermittently consume tank mates as they grow, even with constant prey availability. It can take as long as a few months for twenty to become one.

Individual enclosures for immatures can be made using 5.5 oz cups with lids pierced with two thumbtack holes. Once they reach nearly 1/2″ (~1 cm) body length they can be moved to a 16 oz deli cup or to a 20 oz rectangular

Dendrothereua homa from Arizona, USA.
© John Skierski

Dendrothereua sp., Texas, USA.
© John Skierski

Dendrothereua homa from Arizona, USA (encountered in Tucson, Santa Catalina, and Madera areas: Pima County).
© John Skierski

container with six to ten thumbtack pinholes in the lid. They cannot escape during feeding as long as the sides are kept free of debris, but a minimally vented lid (a few tack holes) is necessary to slow down drying out of the substrate. A quarter inch layer (6 mm) of any soil-like material can be used as substrate to maintain humidity. The consistency is not important since they do not burrow. One side can be sprayed and there should always be a damp area. Frequency of spraying varies greatly with temperature, airflow, ventilation, and watering volume, but the trick is to make sure there is always a wet area. Cage decorations are unnecessary, but a piece of paper or dead leaf can serve as a hiding spot. They seem to prefer to hide and climb, but fare just as well on an exposed horizontal plane from hatchling to adult.

Temperature and lighting within the comfortable range for humans is probably acceptable for any scutigeromorph. Captive specimens have been reared at temperatures between 70° F and 85° F (21° to 29° C). Temperate species especially are capable of surviving well below that range, but hibernation is unnecessary in captivity. Immatures and adults of a few species have been maintained under indirect lighting with sixteen-hour days, though any day length is probably acceptable.

SELECTED SPECIES OF SCUTIGEROMORPHA

Dendrothereua homa (Chamberlin, 1942)

ARIZONA HOUSE CENTIPEDE

Indications: Antennomeres longer than wide. Coloration: Golden overall with yellow around the spiracles. Length: Up to 4/5" (2 cm). Locality: Arizona

This native Arizona species was unheard of a few years ago, but seems to be a rather common inhabitant in the state. They can be found around rocks and are easiest to find at night with a flashlight or near lights where they are attracted to the gathering prey. Specimens have been reared to maturity from eggs and are kept similarly to *S. coleoptrata*. Of course they are not nearly so cold tolerant. Despite the delicate yellow colors, like other scutigeromorphs, this species can transform to a deep purple (followed by black or brown) following death as a result of humidity and decomposition. Its taxonomic identity (Mercurio 2010) is uncertain and needs study and confirmation.

Dendrothereua sp., Texas (Austin outskirts: Travis County). © John Skierski

Dendrothereua sp.

TEXAS HOUSE CENTIPEDE

Indications: Antennomeres much longer than wide, after articulated joint. Coloration: Grayish tergites with a reddish circle around each spiracle. Pale gray legs with little to no banding.

Dendrothereua sp., Texas. © John Skierski

Malaysia scutigerid. © Chia How

Scutigera coleoptrata face.

At the 13-leg-pair stage, immatures only have seven tergites and are rather small. *Scutigera coleoptrata*.

Length: Less than 4/5″ (<2 cm). Locality: Central Texas.

This handsome little species may be common in some areas of Texas, though most animals found in the state are the widespread *S. coleoptrata*. One specimen of this species posted on BugGuide.net was temporarily identified as *Sphendononema guildingii*, a large colorful species found in the West Indies also with red around the spiracles, and later moved to *Scutigera* sp. Although both identifications were seemingly wrong, the final species identification is uncertain. Various spine structures do not match that of *Sphendononema* and the antennomere dimensions of longer than wide are not a feature of the genus *Scutigera*. Though common in some areas of Texas, only a few specimens have been seen in captivity, possibly just males so far, and there are no reproductive reports.

GIANT MALAYSIAN HOUSE CENTIPEDE

Indications: Dark gray with brown, white, and purple highlights. Length: Up to 3″ (75 mm).

This spectacular species was imported in small numbers in 2008 from Malaysia. At least one female produced a number of viable eggs. Survival rate for the immatures was high if the right conditions were met. The minute the humidity dropped, they would desiccate and die. The survival rate, including errors made in drying out some of the young, was about 50% until they were about half-grown. The immatures seemed to be rather hardy beyond husbandry lapses, but at around an inch or better they all just sat down and died. Conditions and diet had remained the same. There is much to learn about the life cycle and rearing of this species, but no livestock to work with. One subadult female matured about two months after arrival and then lived only another seven months. She did not molt again during this time. Her diet was crickets, adult *Blatta lateralis*, juvenile *Blaptica dubia*, grasshoppers, and moths. She would kill prey even when not hungry and would just toss it away (Michael Fanto pers. comm.). Some *Scolopendra* also kill prey when they are not hungry and seem to do this just so the animals stop touching them.

Larger Malaysian scutigerid, 2008, feeding on an adult male *Blatta lateralis*. © Michael Fanto

The specimen in the photographs without banded legs is a different scutigerid collected in Malaysia. It may or may not be in the same genus. The pictured animal was found during the day under a piece of bark while other representatives of this species are usually found under bark or boards lying on the ground during day time. At night specimens may be seen in the open sitting on leaves (Chia How pers. comm.).

Scutigera coleoptrata (Linnaeus, 1758)

HOUSE CENTIPEDE

Indications: The only species found in cold climate. Members of this genus have very short, wide antennomeres (almost like stacked coins), especially near the articulated joint. Length: 1″ to 1.2″ (2.5-3 cm). Coloration: Cream or tan overall with three purplish-brown to black longitudinal stripes and banded legs. Locality: Human structures worldwide, originally from the Mediterranean. This is the only widespread Scutigeromorpha found in North America; adventive.

Scutigerid on a tree in the rainforest Panay, Philippines. © Christian Schwarz

This was the first species described and contrary to some popular texts the species name comes from the "sheathed" tergites, not any resemblance to beetles (beetles or coleoptera have "sheathed" flying wings). The species name could refer to any Scutigeromorph. Many of the oldest names refer to common features of entire groups since they were the only representatives known during Linnaeus' time.

Scutigera coleoptrata can be extremely active and fast. It catches tiny moths and crickets nearly as fast as the eye can register, but runs from too large of food. The house centipede seldom eats low profile foods like isopods, earthworms, or mealworms. It does not burrow, so substrate or a moist paper towel is needed only to retain humidity. The early instars are likewise super-fast and very cannibalistic. Small ones molt every two weeks. In the "wild," time to maturity is usually two years, but captive specimens kept at >70° F (>21° C) and fed small crickets and cockroaches weekly after reaching the 30-leg stage can be sexually mature in twelve months.

The creature most commonly known as the house centipede is the only centipede that has been able to follow man nearly everywhere across the globe. (*Lithobius forficatus* has an impressive man-made range, but it is found in outdoor landscaping and rapidly desiccates in a home.) The house centipede is found in buildings and homes even in the cooler areas of Europe and North America. It ventures outside during the warm parts of the year and hibernates inside rotten logs. Despite some inferences that it only survives in heated human structures in colder areas, specimens may be found under rocks in Northern Ohio in midwinter (pers. observ.). It also occurs outdoors in Jersey, Britain (Lee et al. 2007). In temperate climates it hibernates during the winter even indoors (basements are much cooler in winter and food is lacking) but it will remain active and feeding year-round in captivity. Specimens are almost never, if ever, found in the basement before March or after October.

Certainly the house centipede is a very fast-moving creature—at 1 foot, five inches per second (>42 cm/s) it is presently the fastest moving centipede recorded, but if it gets hit with relatively mild impact it curls up and dies. If the impact hits the legs instead of the body, it drops legs in piles. The pile of legs can twitch for minutes (usually about 30 seconds versus 3 seconds for *Lithobius forficatus* autotomized legs). As long as the animal survives initial impact, the legs grow back close to full size at the next molt. It can also drop legs in stressful conditions and sickly animals can lose large numbers of their legs before dying.

It is strange that the house centipede is so strong and delicate at the same time. It must be a tough critter to follow man and exist nearly everywhere, but it is difficult to understand how. Unlike various pest arthropods, there is no part of its life cycle where it is capable of surviving humidity extremes or harsh handling. Immatures are fragile and rapidly die from lack of food or moisture. It never occurs in large numbers; it is rare to find more than one or two in a house at a time, but both sexes are required for reproduction. Without a drought-resistant resting stage or high reproductive rate (the average female does not appear to produce more than sixty eggs in a lifetime), it is hard to understand how it has been able to spread around the globe.

"It does well because it lives in houses, or at least near man, and profits from pest insects, without being overly specialized. And it can be easily transported with cargo. Mechanical resistance is not a prerequisite for becoming invasive, and gigantic egg numbers are only required if losses are high" (C. Schwarz pers. comm.).

ORDER LITHOBIOMORPHA

INDICATIONS: Fifteen leg pairs, often short legs, and alternating large and small tergites.

Stone centipede is the common name for members of the centipede Order Lithobiomorpha. These are usually the most common centipedes in disturbed areas such as landscaping and gardens. Specimens are often found under stones (*litho bio* means "stone life"), but also beneath wood, logs, and trash. Unlike the other three large centipede orders, stone centipedes tend to be similar in color and shape across species and genera, though there is some variation, including pale cave species with elongate legs and antennae. The biggest lithobiomorphs are rarely much more than an inch and there are many far smaller species such as *Nampabius turbator*, a small, less than .3″ (7.5 mm), cave-dwelling species found in Virginia (Crabill 1952). This is the only centipede order known for numerous troglobitic species, about fifty so far described—compare to Scolopendromorpha at two (White and Culver 2012), geophilomorphs with three (Mercurio 2010) and scutigeromorphs zero. While various centipedes may be found in or near caves, troglobitic species do not live in surrounding habitats, have reduced eyes and pigment, and have elongate appendages. Nevertheless, even the species restricted to caves do not look terribly different at a glance.

Pterygotergum svenhedini female from China. © Guanguanguan

Most stone centipedes are the same color overall, but some such as *Lithobius variegatus* have dark and light bands on the last few leg pairs or a stripe down the back like *Pterygotergum svenhedini*. A few rare species are not in the yellowish to reddish brown range. These include a small (less than 0.5″ or 1 cm) species found at the summit of Mauna Kea in Hawaii which is solid black (Howarth and Mull 1992) and another seen on Haleakala Volcano in Hawaii which is translucent white (Clausen 2008). The 1,120 Lithobiomorpha species are found in just two families:

Family Henicopidae
Family Lithobiidae

Nine out of ten stone centipede species, those from the mostly temperate Family

Lithobiidae, have a pair of "eye patches" or ocular fields composed of multiple ocelli. The number is variable by species and developmental stage, but can total up to at least sixty-eight ocelli (McGavin 2000). These eyes resemble the ocelli clusters of most of the common millipeds and the majority of terrestrial isopods. The remaining ten percent of stone centipede species in the Family Henicopidae are often tropical in distribution and have a single pair of large ocelli. Troglobitic species can be eyeless and this is independent of the family.

Stone centipedes look similar to, and are often confused with, small species from the Order Scolopendromorpha. (In the course of research I noticed a major entomological website with a photo of a scolopendromorph *Cryptops* labeled as *Lithobius forficatus* which was promptly corrected after my suggestion.) They are, however, markedly different. Stone centipedes have physical and developmental features similar to the house centipedes (Scutigeromorpha spp.). Dried stone centipedes contract and often appear to have just eight visible upper plates or tergites like a scutigeromorph. Likewise they hatch from individually laid, thick-shelled eggs, into mobile first instars, have uneven tergite size, Tömösvary organs near the base of the antennae, somewhat complex eyes, and develop fifteen leg pairs only following a set number of larval molts. Scolopendromorphs and geophilomorphs have twenty-one leg pairs at the very least, develop through an immobile protonymph stage, begin life with a full complement of legs, possess tergites similar in size that do not overlap, and most have from zero eyes to a maximum of eight ocelli. Lithobiomorphs, like the Scutigeromorpha, only have spiracles on the segments with enlarged tergites, but there is one on both sides of each segment (not one in the middle) and the tracheae provide all the oxygen to the cells. Spiracles are on the exact same segments in Craterostigmomorpha and Lithobiomorpha, but matching the sternites, not tergites. Craterostigmomorphs are the closest relative, but there could be no confusion unless surveying in New Zealand or Tasmania.

Freshly molted *Lithobius*. © John Skierski

Stone centipedes are harmless to humans and the dwarf species have no ability to deliver venom, but this is because they are so small. The common and larger stone centipedes, like *Lithobius forficatus,* are unable to envenomate most people unless placed against the soft skin of the armpit, and even then the bite has minor effects. As with all centipedes, the "bite" is really a venomous hug since the venom is located in the modified front legs.

Gender determination is easy because the male and female gonopods (terminal segments) are distinct. The females have left and right gonopods that are conspicuous clasping type appendages and usually have spines on them, but the males lack these. Furthermore, the males of some species display notable differences in the structure of the terminal legs. They can have interesting small outgrowths or shape difference of the male's terminals and, when present, can be seen easily under a stereomicroscope.

Bothropolys sp. from China.
© Zhangjunduo

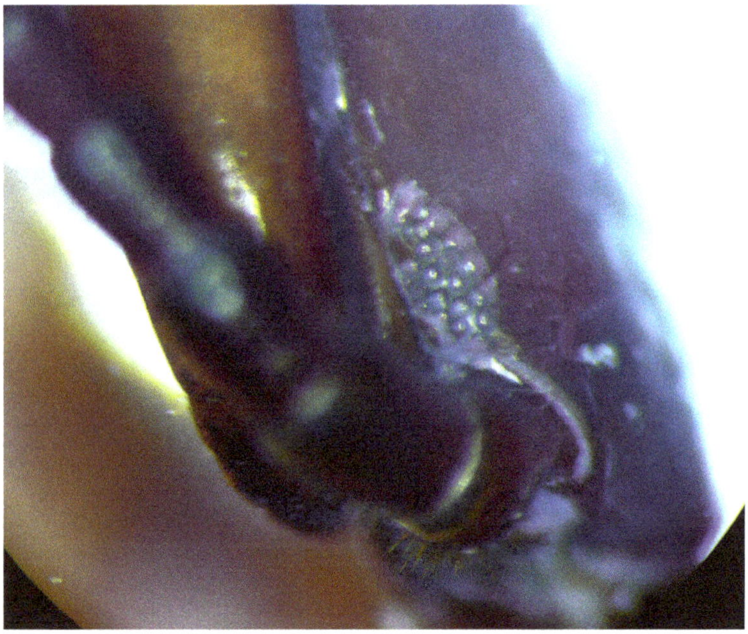

Ocular field of *Bothropolys* sp. from China.
© Zhangjunduo

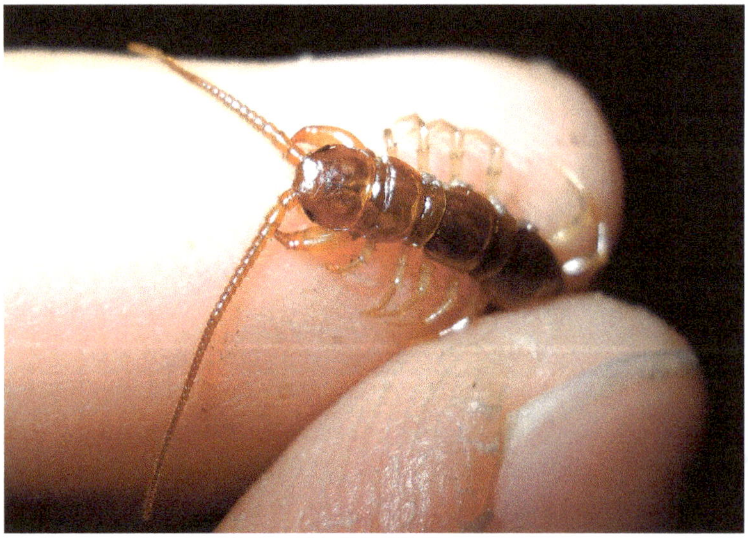

Lithobius forficatus trying to bite.

Tidabius tivius feeding on dry, kibbled dog food.

Egg size compared to the mother, *Tidabius tivius*.

DEVELOPMENT AND CULTURE METHODS

The eggs are tiny and range from 0.004″ to 0.03″ (0.1 mm to 0.8 mm) between the smallest and largest species. Females will lay eggs year-round in captivity and usually bury them individually at least a few millimeters deep in the substrate. Substrates kept very damp show the best results, with dozens of hatchlings found at any time in an active culture. Culture containers in which the substrate is kept lightly damp usually host few to no small immatures. Due to the cryptic nature of oviposition of the tiny eggs and ease of drying out, it is difficult to observe hatching. However, sometimes an egg is seen being carried around at the end of the female's genitalia which can be removed and isolated for hatching. The egg hatches directly into a mobile first instar like scutigeromorphs, but with far more legs (fourteen versus eight). The hatchling is white in color and remains within the chamber it was laid, though it can run just as quickly as later instars. Like house centipedes the 1st instar does not hunt for food, but unlike house centipedes it does eat its own egg shell. After a number of days it molts to 2nd instar, which is the dispersal stage that hunts for food. It gains a single pair of legs and develops a hint of yellow coloration.

The immatures start out pale, nearly white in color and take a number of molts to change to the trademark reddish-tan color of adult stone centipedes. The lesser number of legs would not look strange except the rear legs do not extend far backwards since the final pair is missing from the stumpy terminal segment. They gain four leg pairs at the next molt for a total of twelve pairs, not including the maxillipeds. Following another molt they reach the maximum number of fifteen pairs and have standard looking terminal legs. (Leg counts are 1st instar = 7, 2nd instar = 8, 3rd instar = 12, 4th instar = 15.) The growing centipedes can look like a small adult in form at less than 3 mm, but will not begin to approach adult coloration until growing to 6 mm or better in length.

Culture Container

Plastic shoebox with lid or aquarium with solid glass cover. Very little ventilation is required or suggested.

1″ (2.5 cm) or more of damp substrate such as compost or potting soil.

A number of small pieces of wood or bark.

Add nematodes, springtails, or/and micropods and feed small slivers of fruit every week or two.

Group of dwarf lithobiids, *Tidabius tivius* (Chamberlin, 1909). This is the type species for the rather enigmatic genus *Tidabius*. *Tidabius tivius* specimens have been found in Alabama, Colorado, Illinois, Kentucky, Massachusetts, Michigan, Mississippi, Nebraska, New Jersey, North Carolina, New York, Ohio, Pennsylvania, Tennessee, Utah, Virginia, Washington, and Wisconsin (Mercurio 2010).

Comparison between the adult *Lithobius forficatus* and an adult *Tidabius tivius*.

Leaf litter can improve production and increase hiding places.

UK lithobiomorph. (CC) Mick Talbot

The stone centipedes are very easy to keep if some basic guidelines are met and they are never allowed to dry out. Primarily, it is important to keep the substrate rather damp and feed with fruit and the occasional piece of dog food. It is likely any Lithobiomorpha found in the garden can be reared in the same way. Colonies grow slowly since it can take five months or more for hatchlings to mature. Starvation may result in cannibalism, but just of the dead or dying (it is not just about size—small Scolopendromorpha like *Cryptops* spp. hunt each other). While up to a dozen can be found under one tiny piece of bark, they are almost never touching when initially uncovered. Cannibalism is of course possible and will differ by species, food, and space.

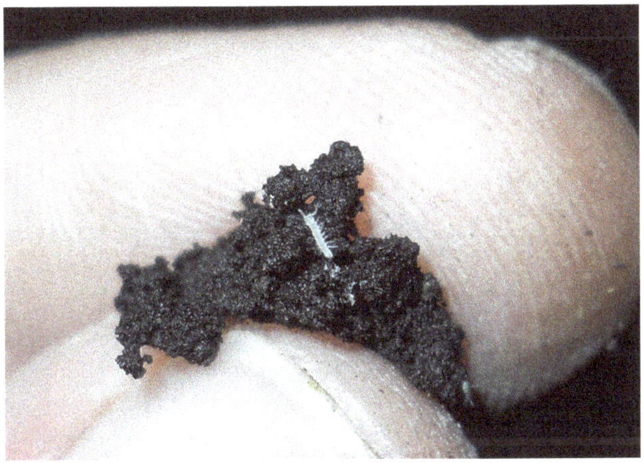
First instar *Tidabius tivius* with seven leg pairs.

Croatian cave-dwelling species, *Eupolybothrus cavernicolus*. (CC) Stoev, et al. 2013.

Lithobiomorphs in the Hobby

Probably the only reason for keeping stone centipedes is to study their activity and development. Few enthusiasts keep them, though these are the animals children like to play with and beginning keeper's often catch a few around the house until they lose interest. Stone centipedes have little value as feeders for other centipedes or predatory arthropods because of time, quantity, and total mass produced. I have used them to feed *Abedus* and waterscorpion immatures in quantity when stone centipedes became established in the wrong cage. Though I have had some issues with them infesting wrong cages, it is mostly because I have not tried to eradicate them. They pose little threat to other arthropod cultures since they rarely jump cages, take time to establish, and are easily eliminated with a substrate change.

SELECTED SPECIES

Lithobius forficatus (Linnaeus, 1758)

The name *forficatus* comes from the Latin term forfex which means *pincer* or *scissors*. This

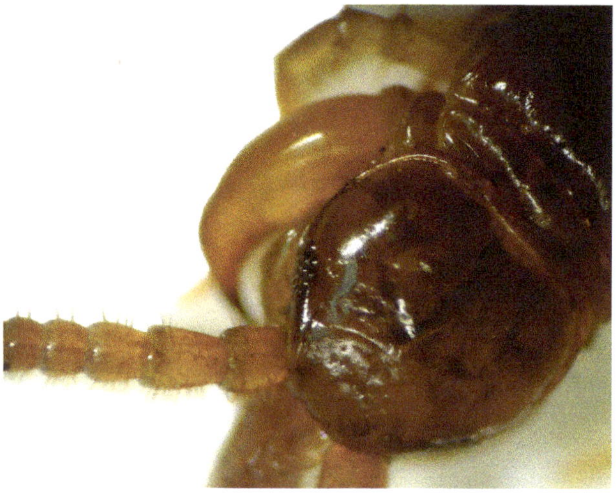

Lithobius forficatus face—the ocular field folds around the side of the head and contains more than thirty ocelli. Note a tiny mite at the rear of the head plate and another on the first tergite.

widespread European species has been introduced to nearly all land masses and is common around disturbed areas, especially small gardens and landscape around the majority of human homes. This is probably one of the most familiar centipedes to anyone reading this book. It is very easy to keep in captivity and may live as long as the longer lived *Scolopendra*. One specimen lived for four years and seven months in captivity and then only died from an accident (Holford 2006). Small specimens feed on fruit, dog food, springtails, and nematodes. Adults eat anything they can overpower and will consume early instar lobster cockroaches with gusto.

Each ocular field contains more than thirty ocelli in adults, but early instars have far fewer eyes. It is difficult to say if ocelli counts reach the maximum when an animal gets to the thirty-leg stage or increase in later instars prior to sexual maturity. It is easy to see the individual ocelli with a 10x loupe, but getting a definitive count at this size is not possible without killing specimens and using a low-powered microscope.

Tidabius tivius

Dwarf Lithobiid ex. Northern Ohio, USA

This stone centipede species appears to be a dwarf since the largest specimens seen over the years have probably never exceeded 3/5" (1.5 cm) in body length. Most adults are only half an inch (1.2 cm), while the smallest egg-laying females have been only 0.9 cm. The very largest are many times less massive than a standard garden *Lithobius*. However, there is a slight chance the size has been limited by the captive environment, since I do not know the size of the original adults. Like other centipedes, they continue to molt after they are capable of sexual reproduction, though adult

Twelve leg-pair (third instar) *Tidabius tivius*.

molts for most species do not result in measurable growth. So far, none of the centipedes have proven to reproduce before approximate adult size, but adult specimens can vary greatly in size by individual or geographic race. This species looks like a small *L. forficatus*, but the head is redder than the body and there are only about eight ocelli that make up each ocular field.

The eggs of this species are white and only 0.004" (1/10 mm) in diameter. Despite the tiny size they can be picked up between two fingers without damage like Scutigeromorpha eggs. Although this is not a terribly fast-growing species, taking six months or more to mature, the eggs take only five days to hatch. The 14-legged first instar molts to the next instar within three days. It is 0.05″ (<1.2 mm) and almost seems too large to have emerged from the tiny egg. The 16-legged 2nd instar is still barely 0.06″ (<1.5 mm) in length and a fraction of that in width. Another three to four weeks and the 2nd instar molts to a 24-legged third instar. Adults do not seem to live much past two years.

I have produced thousands of these dwarf stone centipedes over the last seven plus years. Most of these have not been on purpose since they make their way into giant milliped cages and pose a possible threat to tiny offspring. Still, this has given me the opportunity to observe how they develop under different conditions. I originally introduced some from wood or leaves collected outdoors, and, sometimes, huge colonies have developed over time in certain milliped and isopod cultures. When there are hundreds there seems to be greater die off of the millipeds or isopods, but I have not yet seen evidence of predation, while only the adult dwarf lithobiid could attack a tiny spirobolid milliped. I am not 100% certain they eat healthy specimens of the larger arthropods even in the earliest instars. They probably feed on nematodes and fungus gnat larvae, but observed feeding has been almost entirely on fruit. When medium to large groups have been isolated and offered white micropods (*Trichorhina tomentosa*), various springtail species, and small julid millipeds as prey, the centipedes have slowly withered away over half a year or so. The listed prey do not appear to decline in number as a result of predation. The dwarf stone centipedes may feed on small prey but only as part of the diet. I have learned through observations that they eat fruit and, less enthusiastically, dog food (corn and meal based). Rotting vegetable matter likely is a large part of the diet outdoors. In the case of contaminants, those living with creatures fed primarily freshwater algae and dog food result in low growth compared to those fed watermelon and cantaloupe.

Pterygotergum svenhedini adult pair (female in focus). © Guanguanguan

Pterygotergum svenhedini Verhoeff, 1939
Indications: Male with enlarged tergite (tenth leg pair).

This is the largest and the most fascinating stone centipede in China, though it still does not exceed 1 1/2″ (<4 cm) in body length. In China it is called wings stone centipede or "plane tail" on account of its triangular bottom (Zhangjunduo pers. comm.). The female is similar to the male, but she does not have the expanded tergite.

Males of this species have an unusually shaped twelfth tergite. *Pterygotergum svenhedini*, the largest and the most fascinating stone centipede in China. © Guanguanguan

ORDER GEOPHILOMORPHA

INDICATIONS: Twenty-seven to one hundred and ninety-one leg pairs.

The soil or worm centipedes comprise the least commonly seen order in captivity, though they are the most variable and unusual order. Some species have tiny maxillipeds that are barely noticeable, while others have a huge set which makes them look like a tiny Chinese dragon. They cannot compete with the coloration of Scolopendromorpha, but there are species that are white, yellow, black, or fire engine red. They live in the soil of terrestrial environments, but are also common to areas between high and low tide. Some genera like *Hydroschendyla* are restricted to the intertidal zone, while others can live on the seashore as well as woodland habitats far from the ocean. Many of the littoral species can survive underwater for a few days. Unlike many earth worms, all worm centipedes are without eyes. However, surprisingly, some produce light through chemical secretions (bioluminescence) that must be a warning to predators (Nichols et al. 1971). Glowing representatives are found worldwide and are sometimes known as glowworms, not to be confused with the millipedeating, larviform beetles known as glowworms (Family Phengodidae). Geophilomorphs generally do not exist in arid areas where the soil or sand is too dry. They are also unrepresented in damp caves where there is little to no soil. There are presently thirteen families (Zhang 2011).

This is a male *Arenophilus bipuncticeps* (Wood, 1862), from Medina County, Ohio, USA.

Family Aphilodontidae
Family Ballophilidae
Family Dignathodontidae
Family Eriphantidae
Family Geophilidae
Family Gonibregmatidae
Family Himantariidae
Family Linotaeniidae
Family Macronicophilidae
Family Mecistocephalidae
Family Oryidae
Family Schendylidae

This is a male *Geophilus vittatus* (Rafinesque, 1820), Quebec, Canada. This is one of the species that secretes bioluminescent fluids when it is disturbed sufficiently. It gives off "a faint but distinct blue-green glow that lasts for some seconds" (Jones et al., 1976). © Martin Gamache

Geophilomorphs are most commonly known as soil or worm centipedes (rarely earth, wire, or snail centipedes). The name soil centipede is self-explanatory. The name worm centipede comes from the long flexible body which flexes and looks similar to an earthworm, not because they eat worms. It is very difficult to get even the very largest species to feed on worms in captivity. Many species are red or pink in color and resemble the color of worms as much as shape and movement.

Although the tiny legs can move faster than the eye can register, the animals themselves cannot run away with any speed on flat surfaces. However, they can rapidly yank sections of the body away from a disturbance in the manner of an earth worm. The legs are easy to see but are no bigger than the setae of some marine worms. A few worm centipedes are truly marine, such as *Hydroschendyla submarina* and *Strigamia maritima,* but they are shoreline species that would die in a marine aquarium after more than a few days submergence. Species capable of surviving a period of time below the water's surface are not restricted to the seashore; *Mixophilus indicus* have been found under rocks at the bottom of the Coom River in India along with polychaete worms (Cloudsley-Thompson 1958). Large polychaete worms known as bristle worms are extremely common in live rock and saltwater aquarium gravel. In these worms the spines (all of them) are usually venomous as they penetrate and break off in human skin like the small, irritating barbs of prickly pear

Venezuelides hasei from Peru. Some specimens of this species have exceeded 10″ (>25 cm) according to photographs containing measuring devices. © Paul Bolinger

cactus (*Opuntia* spp.), and they may also cause allergic reactions. Bristle worms are commonly confused with centipedes by the novice, but the "legs" are just inarticulate bristles, and there are no maxillipeds.

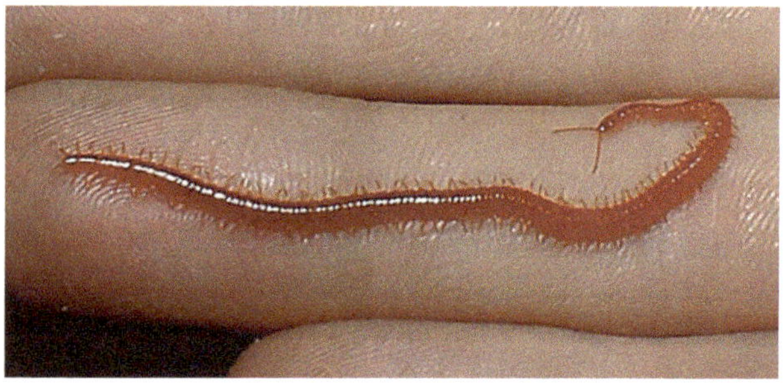

This is a female, most likely *Strigamia bidens* Wood, 1862, from Alabama, USA.

The reason for the limited interest of hobbyists is even a 4″ (10 cm) specimen is a relatively tiny animal and smaller animals can look more like a piece of string than a creature. The only truly impressive species I am aware of is a giant black geophilomorph from South America that reaches 10″ (25 cm) in length, *Venezuelides hasei*. The specimen pictured here was collected in 2010 in Peru, but the Peruvian exporter lost his export license shortly before shipment. This incredibly long species may rival *Scolopendra gigantea* in total trunk length, but not even remotely in mass.

These are the only centipedes that commonly have leg counts close to 100. (The name centipede means *100 feet* but when the maxillipeds are not counted, the closest have either 98 or 102.) Males typically have fewer legs than females. Nevertheless, this is the only order in which the number of legs (and corresponding segment number) is somewhat variable among individuals of the same species and can be unrelated to gender. Half a dozen species vary geographically, with more northerly specimens having fewer body segments (Arthur and Kettle 2001). Leg counts appear to be slightly variable even for the same individual of a few species, although changes in leg counts with

Female *Strigamia parviceps* Wood, 1862 from Klamath County, Oregon, USA. © John Skierski

development is not a rare feature. It is however rare for epimorphic centipedes since epimorph means the form does not change. The only reason some species are believed to gain a few leg pairs is through counting and comparing average number of legs for second instars and wild adult specimens (Lewis 1981).

The long, flexible body and short legs allow them to move through loose dirt and small tunnels in the substrate with ease. They often fold their body back and forth so one animal at rest can look like a tiny pile of spaghetti. It is difficult to imagine what benefit such a long, cumbersome body provides, but this form must have its advantages considering the large number of very successful species. The body itself is functional as an appendage in its own right. Not only can it be used to cross gaps, but it also plays a part in defensive displays. The body can fold up and unfold like an accordion for rapid movement. They can fold and unfold as they run backwards in flight mode.

Geophilomorphs can detect light even though they do not have eyes. "I have observed this with my own specimens. I would walk down the stairs late at night and not turn the light on until I got to the bottom of the stairs where I had geophilids in petri dishes. At first they would be on the surface of the substrate but as soon as I turned the light on they would bury!" (Mercurio pers. comm.).

Many species produce defensive fluids when molested which may ward off small attackers and provide a foul taste in the mouth to larger predators. As mentioned, more than a few species also produce a temporary glow or flash of light when the secretion is excreted. The light may simply be a byproduct of the defensive chemicals contacting the air or mixing together. Bioluminescence is usually a product of two different secretions coming in contact with each other (containing luciferin and luciferase) and does require oxygen (Anderson 1980). Still, most terrestrial animals are startled by spontaneous light. *Scolioplanes crassipes* from the Vosges in France turns bright green when excited (Stanek 1969).

Pairs are sometimes found wrapped up together under a rock. Reproduction can be very similar to the Scolopendromorpha with advanced maternal care through second instar, but some species simply lay a pile of eggs. These would be much easier to keep since brooding

Geophilomorpha can have highly adapted terminal legs like the other orders, Thurston County, Washington. "This is a male, most likely *Strigamia cephalica* Wood, 1862; this is a new state record if I am right and this animal has not been seen since it was described as far as I know. This is why you have to preserve stuff like this after it dies!"(Mercurio pers. comm.) © John Skierski.

For this female *Strigamia fulva* Sager, 1856, the rear legs look very much like the rest. Male geophilomorpha terminals are often inflated. Pierce County, Washington, USA. © John Skierski

females should not be disturbed or they may eat their offspring, like scolopendromorphs.

Arenophilus bipuncticeps

This species pictured from northern Ohio is yellow with a reddish head and adults are 1 1/2″ (3.7 cm) long. I kept two adults in a 16 oz deli-cup with a little less than 0.5″ (1 cm) of substrate for nine months and then released. They were offered tiny captive-hatched stone centipede, julid millipeds, and springtails without any notable feeding seen.

Strigamia sp.

Some of the specimens pictured here were collected from near Klamath Falls, Oregon, and housed together without cannibalism. They would not eat every time food was offered, but would feed on freshly killed roach nymphs now and then (Skierski pers. comm.).

ORDER CRATEROSTIGMOMORPHA

INDICATIONS: Fifteen leg pairs and twenty-one tergites. One pair of eyes.

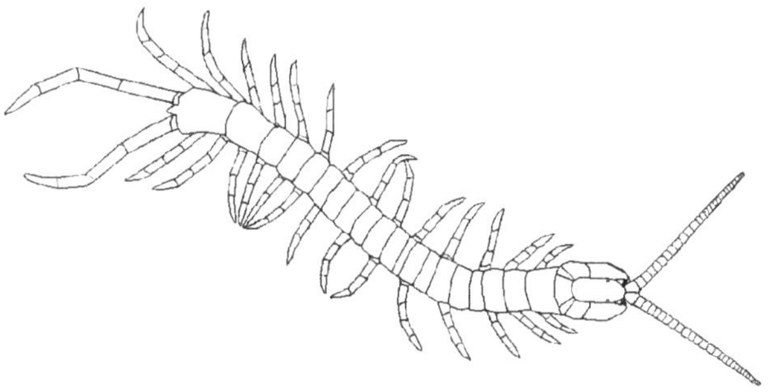

For one hundred and six years this unusual order with a single family and genus was comprised of a single species thought to be widespread on New Zealand (North, South, and Stewart islands), as well as the large Australian island of Tasmania. However recent molecular analysis has shown the Tasmanian and New Zealand natives represent unique species. There is of course more to the story since it was not always considered a monotypic order. A few decades after Pocock created the order, it was reduced to a suborder of Lithobiomorpha by Attems (1926) since, among other similarities, the spiracles are located on the same body segments, the maxillipeds are hinged alike, and there is a structure where the Tömösvary organ should be located. Since that time there has been some debate. Manton (1965) thought *Craterostigmus* should fit in with the Scolopendromorpha, but the very few similarities in body structure are superficial and may be coincidental or relate to adaptations for brood care. The close relation to Lithobiomorpha is not in question for the vast majority of published works, and the only hint of debate is whether or not it deserves order or suborder status.

Family Craterostigmidae
- Genus *Craterostigmus*
  - *Craterostigmus crabilli* Edgecombe and Giribet, 2008
  - *Craterostigmus tasmanianus* Pocock, 1902

These species have a single ocellus on each side of the head. The eye is an intricately lobed structure which is far more complex than any large ocelli of stone centipedes so far examined, lending further support to the order designation (Müller and Meyer-Rochow 2006).

Craterostigmus are relatively small animals, hardly more impressive than the largest common *Lithobius* sp. stone centipedes. The New Zealand species (*C. crabilli*) grows to a maximum of 1.5" (3.7 cm) while the Tasmanian animal (*C. tasmanianus*) can be smaller, but is supposed to grow to nearly 2" (3 to 5 cm). Other than possible difference in size, the only way to tell the two species apart is through genetic analysis or collection location.

Wild observations of *C. tasmanianus* suggest it produces one brood of young and watches over them like *Scolopendra*, but for a much longer period of time. In the wilds of Tasmania, females have been found guarding the eggs in late September and holding young in April (Mesibov 1995). To date (January 2014) there is no documentation of the early developmental stages. (Sources claiming the immatures begin with twelve legs are referencing the free-living young found far from the adult females after the six-month brooding period.) It seems likely they molt once or twice before leaving the mother because no other centipede hatches directly to a roaming predacious stage. (First instar Lithobiomorpha and Scutigeromorpha are mobile but do not hunt, and molt within days to the next instar.) While it is possible *Craterostigmus* sit in a single undeveloping developmental stage the entire brooding period, considering the basic similarities to stone centipedes, it seems just as reasonable to suggest they molt twice in their mother's care as not at all. Factors hindering knowledge of their life cycle include lack of interest in studying these species, unavailable livestock, and the likelihood females eat young if disturbed. Juveniles are often found in the leaf litter at a stage where they have twelve leg pairs, like third instar lithobiomorphs. Brood sizes range from forty-four to seventy-seven (Mesibov 1995). Brood size, times, and early instar observations have not been made for the New Zealand species.

Although collecting in New Zealand is strictly regulated, the species from New Zealand has a chance of being exported one day since it is common and widespread, and because velvetworms (onychophorans) from New Zealand are seen every so often on vendor lists. However, since this centipede is the same color and size of the ubiquitous stone centipedes, I am one of the only humans on earth who would pay a decent price to get a live one, and then honestly out of novelty as much as an interest in observing early development. The massively elongate maxillipeds, odd tergites, and unusually structured anal-genital capsule are interesting, but hardly differentiate them from an abnormal stone centipede, at least in photos.

CLOSING

Centipedes can be kept as spectacular and hardy pets. They grow well on live foods that are easy to purchase, collect, or culture. Thoughtful caging design prevents escape and offers the centipede a long life. New species or color varieties continue to enter the hobby while captive breeding–hatching efforts provide higher numbers of quality animals each year. Possibly even the giant Scutigeromorpha and Geophilomorpha will become accessible as the variety of centipedes available to the chilopod enthusiast grows. Hopefully the photos and information presented here will ignite the passion of those considering keeping these incredible creatures. There is great value in humans understanding and appreciating the natural world directly and this includes the marvelous centipedes, since scientific study and preservation require knowledge and understanding. (Appreciation is a more realistic source for funding than uninformed altruism.)

Australian House Centipede, *Allothereua maculata*. (CC) Chris Birchill

Scolopendromorph, Chitwan National Park, Nepal. (CC) John Pavelka

Banded kingfisher with prey. Thailand. © Panu Ruangjan

GLOSSARY

Adventive: Naturalized or introduced species.

Anamorphic: Development in which larvae hatch with a reduced number of legs and segments that are added as they progress through molts until reaching a set maximum number (30 for all anamorphic centipedes). Later molts result in no more legs. There are no initial, immobile protonymph or postembryo stages.

Antennomere: An article or small segment that makes up the antenna; antennal article.

Autotomy: Reflexive release of an appendage at a specific point of attachment in order to facilitate escape from a predator or to regenerate a damaged appendage. The body part may twitch or writhe after separation to provide distraction. Common for centipede, spider, and phasmid legs, or lizard and earthworm tails.

Brood: (noun.) Litter, offspring, or progeny. (verb.) To incubate.

Coxa: Basal segment of the leg or maxilla.

Coxopleuron: Area of the terminal segment where the base of each leg attaches to a small tergite and greatly reduced pleurites.

Coxosternite: Enlarged, paired, fused coxae of the maxillipeds.

Coxosternum: The large plate below the head made up of the bases of the legs fused to the tergite of the maxilliped body segment.

Desiccate: To dry up or dry out; dehydrate to the point of death.

Dimorphic: Possessing two distinct forms or shapes. Sexually dimorphic males and females are visibly different in shape. This may be expressed in the terminal segment or legs of centipedes.

Dorsal: Top or back.

Endemic: Native to a geographic area.

Epimorphic: Development in which larvae hatch from the eggs with a complete (rarely nearly complete) complement of legs and body segments.

Exocuticle: Outer layer of the exoskeleton.

Extant: Living or presently in existence.

Exuvium: The shed exoskeleton.

Femur (*pl.* Femora): Segment of the centipede leg between the prefemur and tibia.

Frass: The solid waste of arthropods.

Head Capsule: Dorsal section of the head holding the eyes and antennae.

Holotype: A single specimen used as the basis for a species description and name.

Hypopus: A unique stage of certain types of mites that is a response to reduced food, they molt into a suction cup stage and search out exoskeletons of living arthropods to cling to; phoretic stage.

Instar: Standard developmental stages of an immature arthropod terminating in a molt and prior to sexual maturity.

Intersegmental Membrane: Thin, stretchable material such as that connecting leg segments

Scolopocryptops sp. caught in Zhejiang province, China. © Zhangjunduo

and tergites to each other, as different from the pleurite.

LAMELLATE: Having expanded and flattened extensions. (In centipedes this refers to the flattened teeth seen on the second maxillary palps of certain species like *Scolopendra heros* and *Scolopendra gigantea*, whereas the common entomological usage refers to the antenna shape of certain beetles.)

LITTORAL: Relating to the area between high and low tide.

MANDIBLES: Central chewing pair of mouthparts; jaws.

MAXILLAE: Paired mouthparts located behind and outside of the mandibles, basal segments usually end in a sharp point and function like pincers while the extremities form maxillary palps.

MAXILLARY PALPS: Terminal, finger-like, manipulative appendages formed from the outer segments of the maxillae.

MAXILLIPEDS: Highly specialized front legs of centipedes used like mouthparts and usually able to inject prey with venom (maxillipeds are highly specialized and not included when counting legs); prehensors; forcipules; outer, fourth pair of "mouthparts" that arise from the first trunk segment of the body.

MARGIN: Border or perimeter; in most invertebrates this refers to coloration bordering a segment but in centipedes it more commonly refers to a raised ridge on the edge or border of the tergites.

OCELLUS (*pl.* OCELLI): Light detection organs.

OMMATIDIUM (*pl.* OMMATIDIA): Structural unit or complete facet of the compound eye.

OVIPOSITOR: A tubular structure through which a female centipede lays eggs.

PALP: Short for palpus; elongate segmented appendage normally used to taste and manipulate food, located near the mouth of an arthropod.

PARAMEDIAN: Near the middle.

Parthenogenesis: Development of the egg without fertilization.

PHORETIC: Describing an animal which attaches to another animal for transportation.

PLEURAL REGION: An area of flexible integument along the sides of the centipede; between the sternites and tergites.

PLEURITE: One section of stretchable, lightly sclerotized exoskeletal material on the side of a body segment between a tergite and sternite.

POSTULTIMATE MOLT: A molt that takes place after sexual maturity has been reached and therefore after the ultimate molt.

PREFEMUR (*pl.* PREFEMORA): Segment of the leg between the trochanter and femur.

Long-legged Centipede, *Thereuopoda clunifera*, Tai Mo Shan, Hong Kong. (CC) Thomas Brown.

PREMOLT: Period preceding the molt and usually signified by changes in behavior or color.

PROCESS: A large exoskeletal outgrowth on which smaller spines are located, primarily used for the large inward facing spur on the terminal prefemur.

PROTONYMPH: Immobile developmental stage in epimorphic centipedes between egg and first instar that resembles a croissant or tire; second embryonic instar.

PROXIMAL: Closest to the point of attachment.

PROXIMOTARSAL SPUR: Thickened ventral spine on the proximal (first or basal) segment of the tarsus.

RING FURROW: Deep, crescent-shape groove that runs across the front of the first tergite; common for New World species. Also known as the anterior transverse sulcus.

SCLEROTIZE: To cause rigidity of the exoskeleton through crosslinking of the proteins to form sclerotin, which may be of various forms and a substance other than chitin.

SETA (*pl.* SETAE): Hollow growth from the epidermal layer of the body wall that are often sensory in function. Hair-like setae are often referred to as "hair" but are principally different from the hair of mammals.

SEXUALLY DIMORPHIC: Presence of noticeably different external gender characteristics.

SPERMATOPHORE: A specialized sperm packet.

SPIRACLES: Air exchange holes located on every second or third trunk segment on each side of the body in the pleural region (located singularly at the rear of scutigeromorph tergites).

SPUR: Outgrowth of the exoskeleton, often larger than a spine and may have one or more spines at the end, but often used interchangeably with spine.

STERNITE: The rigid ventral surface or plate located on each segment on the trunk of a centipede.

STRIDULATION: The act of making noise by rubbing body parts together.

SUTURE: A visible, shallow indentation running across part of a body plate; groove; sulcus.

TARSOMERE: Small segments that form the tarsus or foot.

TARSUNGULUM: Large claw at the end of the maxilliped.

TARSUS: The foot, or ultimate major division of the arthropod leg.

TENERAL: Soft, often pale or white-colored state of the arthropod exoskeleton immediately following a molt.

TERGITE (*sing.* noun): A thickened dorsal plate covering the trunk segments of a centipede's body. (Tergal, adj.) of the tergite.

TERMINALS: The ultimate or rear legs of a centipede; prehensorial anal legs.

THIGMOTACTIC: A preference to feel surfaces against the top and bottom of the body.

TIBIA: Leg segment between the femur and tarsus.

TÖMÖSVARY ORGAN: Advanced sensory pit located above the base of the antenna which may serve multiple functions including sound, pressure, odor, and humidity detection; found on anamorphic centipedes, millipeds, and some other "primitive" arthro-pods.

TROGLOBITIC: Being found exclusively within and physically adapted for life inside caves.

TROCHANTER: Leg segment between the coxa and prefemur of the centipede leg.

ULTIMATE MOLT: The final shedding of the exoskeleton from a sexually immature arthropod in resulting in a mature specimen capable of reproduction.

VENTRAL: The underside or belly.

VICARIANT: To be separated (for populations of organisms) by a geographic barrier such as a mountain.

Scolopendromorpha are not entirely unique among centipedes in feeding on vertebrates in nature. This wild Scutigeromorpha in Borneo has just caught a small frog. © Nature`s Canvas

BIBLIOGRAPHY

Anderson, D. T. (2008) Mating *Scolopendra heros*. *Invertebrates Magazine* 7(2): 2-6.

Anderson, D. T., and O. McMonigle. (2013) Development and husbandry considerations for the Indian tiger centipede *Scolopendra hardwickei* Newport, 1844. *Invertebrates Magazine* 13(1): 4-8.

Anderson, J. M. (1980) Biochemistry of centipede bioluminescence. *Photochemistry and Photobiology* 31(2): 179-181.

Arthur, Wallace, and Chris Kettle. (2001) Geographic patterning of variation in segment number in geophilomorph centipedes: clines and speciation. *Evolution & Development* 3(1): 34-40.

Attems, C. (1930) Myriapoda 2. Scolopendromorpha. *Das Tierreich* 54: 1-308.

Babb, R. D., and C. E. Busboom. (2013) *Pituophis catenifer affinis* (Sonoran gophersnake). Predation. Natural History Notes. *Herpetological Review* 44(4): 696.

Baerg, W. J. (1924) The effect of the venom of some supposedly poisonous arthropods. *Annals of the Entomological Society of America* 17(3): 343-352.

Baxter, Ronald N. (1993) *Keeping and Breeding Tarantulas*. Chudleigh, Ilford, Essex.

Bonato, L., Edgecombe, G. D., and M. Zapparoli. (2011) Chilopoda—Taxonomic overview. In: Minelli, A., ed. *Treatise on Zoology—The Myriapoda*. Volume 1. Brill, Netherlands.

Bücherl, Wolfgang. (1971) *Venomous Animals and their Venoms*. Vol. 3. *Venomous Invertebrates*. Academic Press, London.

Breene, R. G., III. (1997) Sinister centipedes. *ATS Forum* 6(6): 183-187.

Bruins, Eugene. (1999) *Encyclopedia of Terrarium*. Rebo Productions, Minkowsky, Netherlands.

Butler, Henry D. (1856) *Family Aquarium: or Aqua Vivarium*. Dick & Fitzgerald Publ., New York, NY.

Chipman, Ariel D., Arthur, Wallace, and Michael Akram. (2004) Early development and segment formation in the centipede, *Strigamia maritima* (Geophilomorpha). *Evolution & Development*. 6(2): 78-89.

Clausen, Peter. (2008) Paradise! . . . But where's all the bugs? Part II. *Invertebrates-Magazine* 7(2): 19-21.

Cloudsley-Thompson, J. L. (1952) Collecting centipedes and millipedes. *Bulletin of the Amateur Entomological Society* 11: 5-8.

Cloudsley-Thompson, J. L. (1958) *Spiders, Scorpions, Centipedes and Mites*. Pergamon Press, Oxford.

Crabhill, Ralph E., Jr. (1952) A new cavernicolous *Nampabius* with a key to its Northeastern American congeners (Chilopoda: Lithobiidae). *Entomologica News*. 63: 203-206.

Ditzel, Nicholas. (2002) Observations on reproduction and considerations on the taxonomy

(CC) Oregon Caves National Monument (NPS)

Lithobiidae, Galicia, Spain. (CC) N. Feans

of Asian centipedes. *Invertebrate* (Fall) [Roskilde, Denmark]. (Figure labeled "*Scolopendra subspinipes japonica*" is certainly *Otostigmus*, probably *O. scaber*.)

Dunn, Gary A., ed. (1993) *Caring for Insect Livestock: An Insect Rearing Manual*. Special Pub. No. 8. YES, Lansing, MI.

Edgecombe, G. D., and G. Giribet. (2006) A century later—a total evidence re-evaluation of the phylogeny of scutigeromorph centipedes (Myriapoda: Chilopoda). *Invertebrate Systematics* 20(5): 503-525.

Edgecombe, G. D., and G. Giribet. (2009) Phylogenetics of scutigeromorph centipedes (Myriapoda: Chilopoda) with implications for species delimitation and historical biogeography of the Australian and New Caledonian faunas. *Cladistics* 25: 406-427.

Evans, Arthur V. (2007) *Field Guide to Insects and Spiders of North America*. Sterling Publishing Co., New York, NY.

Edgecombe, Gregory D., and Gozalo Giribet. (2007) Evolutionary biology of centipedes (Myriapoda: Chilopoda). *Annu. Rev. Entomol.* 52: 151-70.

Forster, R. R., and L. M. Forster. (1970) *Small Land Animals of New Zealand*. Revised Ed. John McIndoe Ltd., Dunedin, New Zealand.

Frye, Frederic L. (1992) *Captive Invertebrates: A Guide to Their Biology and Husbandry*. Krieger Publishing Co., Malabar, FL.

Greenaway, Theresa. (2000) *Centipedes and Millipedes*. Raintree Steck-Vaughn Publishers, Austin, Texas.

Harada, K., Asa, K., Imachi, T., Yamaguchi, Y., and K. Yoshida. (1999) Centipede inflicted postmortem injury. *Journal of Forensic Sciences* 44(4): 849-850.

Hase, A. (1928) Neue beobachtungen uber die wirkung der bisse von tasuendfussen (Chilopoda). *Beitrage zur Experimentellen Parasitologie*. 1(1): 76-99.

Holford, Nick. (2006) Collecting centipedes and millipedes. *Bull. Amat. Ent. Soc.* 65(467): 154-161.

Howarth, F. G., and W. P. Mull. (1992) *Hawaiian Insects and Their Kin*. University of Hawaii Press, Honolulu, HI.

Jass, Joan, and Barbara Klausmeier. (2001) *Terrestrial Isopod (Crustacea: Isopoda) Atlas for Canada, Alaska and the Contiguous United States*. Contributions in Biology and Geology, Number 95, October 24, Milwaukee Public Museum, WI.

Jones, T. H., Conner, W. E., Meinwald, J., Eisner, H. E., and T. Eisner. (1976) Benzoyl cyanide and mandelonitrile in the cyanogenetic secretion of a centipede. *J. Chem. Ecol.* 2(4): 421-429.

Kronmüller, C. (2012) Review of the subspecies of *Scolopendra subspinipes* Leach, 1815 with the new description of the South Chinese member of the genus *Scolopendra* Linnaeus, 1758 named *Scolopendra hainanum* spec. nov. (Myriapoda, Chilopoda, Scolopendridae). *Spixiana* 35(1): 19-27.

Lee, P., Barber, T., and S. Gregory. (2007) Collecting centipedes, millipedes and woodlice: the true story. *Bull. Amat. Ent. Soc.* 66(472): 117-123.

Lewis, J. G. E. (1981) *The Biology of Centipedes*. Cambridge University Press, New York, NY.

Lewis, J. G. E (2006) On the scolopendromorph centipede genus *Mimops* Kraepelin, 1903, with a description of a new family (Chilopoda: Scolopendromorpha). *Journal of Natural History* 40: 19-20.

Manton, S. M. (1952) The evolution of arthropod locomotory mechanisms. Part 3. The locomotion of Chilopoda and Pauropoda. *J. Linnean Society (Zoology)* 42: 118-166.

Manton, S. M. (1965) The evolution of arthropodan locomotory mechanisms, Part 8. Functional requirements and body design

Scolopendromorph, Genovesa Island, Galapagos Archipelago, Ecuador. (CC) Alan Davey

California geophilomorph found in bathtub. (CC) Derrick Coetzee

in Chilopoda, together with a comparative account of their skeleto-muscular systems and an Appendix on a comparison between burrowing forces of annelids and chilopods and its bearing upon the evolution of the arthropodan haemocoel. *J. Linnean Society (Zoology)* 46: 251-484.

Mascarino, Pierrino. (1996) Digest of modern invertebrate thought. *Invertebrata* 0(0): 11.

McGavin, George C. (2000) *Insects, Spiders and Other Terrestrial Arthropods*. Dorling Kindersly, New York, NY.

McMonigle, O. (2003) *Giant Centipedes: The Enthusiast's Handbook*. Elytra and Antenna, Brunswick, OH.

McMonigle, O. (2006) Notes on captive reproduction of the feathertail centipede. *Invertebrates Magazine* 5(2): 2-5.

McMonigle, O. (2007) Opdraet affjerhaleskolopenderen i fangenskab. *Exotiske Insekter* 62: 35-39.

McMonigle, O. (2008) Captive breeding Scutigeromorpha centipedes. *Invertebrates-Magazine* 7(2).

McMonigle, O. (2011) *Invertebrates For Exhibition*. Coachwhip Pub., Landisville, PA.

McMonigle, O. (2012a) *Millipeds in Captivity: Diplopodan Husbandry and Reproductive Biology*. Coachwhip Pub., Landisville, PA.

McMonigle, O. (2012b) *The Ultimate Guide to Breeding Beetles: Laboratory Culture Methods*. Coachwhip Pub., Landisville, PA.

McMonigle, O. (2013a) *Breeding the World's Largest Arachnid: Amblypygid Biology and Natural History*. Coachwhip Pub., Greenville, OH.

McMonigle, O. (2013b) Stone centipede culture methods. *Invertebrates-Magazine*. 12(4): 14-16.

McMonigle, O. (2014) Featured invertebrate: *Scolopendra dehaani*. *Invertebrates-Magazine* 13(2).

McMonigle, O., and D. T. Anderson. (2011a) Coloration, subspecies, and captive hybridization of *Scolopendra heros* Girard 1853. Part I. *Invertebrates-Magazine* (March).

McMonigle, O., and D. T. Anderson. (2011b) Coloration, subspecies, and captive hybridization of *Scolopendra heros* Girard 1853. Part II. *Invertebrates-Magazine* (June).

McMonigle, O., and P. Clausen. (2002) *Ghosts of the Trees: Walkingsticks and Leaf-Insects*. Elytra and Antenna, Brunswick, OH.

Mercurio, R. J. (2010) *An Annotated Catalog of Centipedes (Chilopoda) from the United States of America, Canada and Greenland (1758–2008)*. Xlibris, Bloomington, IN.

Mesibov, R. (1995) Distribution and ecology of the centipede *Craterostigmus tasmanianus* Pocock, 1902 (Chilopoda: Craterostigmomorpha: Craterostigmidae) in Tasmania. *The Tasmanian Naturalist* 117: 2-7.

Minell, Alessandro. (2011) *Treatise on Zoology: Anatomy, Taxonomy, Biology. The Myriapoda. Volume 1*. Brill, Netherlands.

Molinari, J., Gutierrez, E., De Ascencao, A. A., Nassar, J. M., Arends, A., and R. Marquez. (2005) Predation by giant centipedes, *Scolopendra gigantea*, on three species of bats in a Venezuelan cave. *Caribbean Journal of Science* 41(2): 340-346.

Müller, C. H., and V. B. Meyer-Rochow. (2006) Fine structural description of the lateral ocellus of *Craterostigmus tasmanianus* Pocock, 1902 (Chilopoda: Craterostigmomorpha) and phylogenetic considerations. *Journal of Morphology* 267(7): 850-865.

Murphy, Frances. (1980) *Keeping Spiders, Insects and Other Land Invertebrates in Captivity*. John Bartholomew & Sons, Edinburgh, Scotland.

Nichols, David, Cooke, John, and Derek Whitely. (1971) *The Oxford Book of Invertebrates*. Oxford Univ. Press, Manchester.

O'Toole, Christopher, ed. (2002) *Firefly Encyclopedia of Insects and Spiders*. Firefly Books, Buffalo, NY.

Sandefer, Carl. (1998a) *The Giant Centipedes of the Genus Scolopendra: Their Captive Care And Husbandry*. Privately printed. [Note: Figure 42 is a banded *S. h. heros* mistakenly identified as *S. h. arizonensis*, though it does not fit the description listed in the publication and certainly came from Texas. Figs. 7 and 12 are *S. h. arizonensis* as labeled. "Scolopendra morsitans" is *Ethmostigmus trigonopodus*.]

Sandefer, Carl. (1998b) Welcome to the *Scolopendra* nightmare. *Webbings* 1(1).

Sandefer, Carl. (1998c) *Scolopendra heros*: The giant centipede of the American Southwest—It's natural history and captive care. *Webbings* 1(2): 3-4.

Serinken M., Erdur, B., Sener, S., Kabay, B., and A. Cevik. (2004) A case of mortal necrotizing fasciitis of the trunk resulting from a centipede (*Scolopendra moritans*) bite. *The Internet Journal of Emergency Medicine* 2(2). http://ispub.com/IJEM/2/2/12582

Shelley, R. M. (2002) *A Synopsis of the North American Centipedes of the Order Scolopendromorpha (Chilopoda)*. Virginia Museum of Natural History, Martinsville, VA.

Shelley, R. M. (2008) Revision of the centipede genus *Hemiscolopendra* Kraepelin, 1903: and possible misidentifications of *Scolopendra* spp.; proposal of *Akymnopellis*, n. gen. and redescriptions of its South American components (Scolopendromorpha: Scolopendridae: Scolopendrinae) *International Journal of Myriapodology* 2: 171-204.

Shelley, R. M., Edwards, G. B., and A. Chagas, Jr. (2005) Introduction of the centipede *Scolopendra morsitans* L., 1758, into northeastern Florida, the first authentic North American record, and a review of its global occurrences (Scolopendromorpha: Scolopendridae: Scolopendrinae). *Entomological News* 116(1): 39–58.

Shelley, R. M., and S. B. Kiser. (2000) Neotype designation and a diagnostic account for the centipede, *Scolopendra gigantea* L. 1758, with an account of *S. galapagoensis* Bollman 1889. *Tropical Zoology* 13: 159-170.

Shelley, R. M., Perreira, W. D., and D. A. Yee. (2014) The centipede *Scolopendra morsitans* L., 1758, new to the Hawaiian fauna, and potential representatives of the "*S. subspinipes* Leach, 1815, complex" (Scolopendromorpha: Scolopendridae: Scolopendrinae). *Insecta Mundi* 2014(0338): 1-4.

Stanek, V. J. (1969) *The Pictorial Encyclopedia of Insects*. Hamlyn House, London.

Stoev, P., et al. (2013) *Eupolybothrus cavernicolus* Komerièki & Stoev sp. n. (Chilopoda: Lithobiomorpha: Lithobiidae). *Biodiversity Data Journal* 1: e1013.

Stoops, E. D., and J. L. Martin. (1995) *Scorpions and Venomous Insects of the Southwest*. Golden West Publishers, Phoenix, AZ.

Supakthanasiri, P., et al. (2004) Anaphylaxis to centipede bite. *Journal Allergy Clin. Immunol.* 113: S244.

Swift, Kelly. (1997) Deep South adventure. *Invertebrata* 1(3): 39-45.

Waldbauer, Gilbert. (1998) *The Handy Bug Answer Book*. Visible Ink Press, Detroit, MI.

Werner, F., and C. Olson. (1994) *Insects of the Southwest*. Fisher Books, Tucson, AZ.

White, W. B., and D. C. Culver. (2012) *Encyclopedia of Caves*. Academic, Waltham, MA.

Wood, Horatio. (1862) On the Chilopoda of North America. *Journal Acad. Nat. Sci. Philadelphia* 2(5): 2-52.

Zhang, Zhi-Qiang, ed. (2011) Animal biodiversity: An outline of higher-level classification and survey of taxonomic richness. *Zootaxa* 3148: 1-237.

A South African scolopendromorph.
(CC) Derek Keats

INFORMATIONAL AND EDITORIAL CONTRIBUTORS

Orin McMonigle has published half a dozen articles on captive husbandry and breeding three of the five chilopod orders, and authored *Giant Centipedes: The Enthusiast's Handbook* (2003).

David T. Anderson is recognized for pioneering work in centipede breeding and hybridization, as well as authoring articles relating his experiences.

Christian Schwarz is an entomologist and tropical ecologist who has worked on dozens of scientific texts, as well as providing editorial assistance on arthropod books.

John Skierski has maintained a large number of chilopod species and has reproduced various *Scolopendra* species in captivity.

Randy Mercurio studies all aspects of centipedes in his spare time, including their natural history, biology, taxonomy, biogeography, and venom.

Annie Lastar has been assistant editor for *Invertebrates-Magazine* (2001-2014) and has edited a variety of other texts.

Tammy Wolfe is a nature photographer and text editor.

Rowland Shelley is a centipede taxonomist specializing in Scolopendromorpha.

Wade Harrell is has been keeping various centipedes as pets for fifteen years and is the former president (and current vice-president) of the American Tarantula Society. He is the curator of live animals at the Harrell House Bug Museum in Santa Fe, New Mexico.

MILLIPEDS IN CAPTIVITY
ORIN A. MCMONIGLE

PILLBUGS AND
OTHER ISOPODS
CULTIVATING VIVARIUM CLEAN-UP CREWS AND
FEEDERS FOR DART FROGS, ARACHNIDS, AND INSECTS
Orin McMonigle

THE ULTIMATE GUIDE
TO BREEDING BEETLES
COLEOPTERA LABORATORY CULTURE METHODS
ORIN MCMONIGLE

KEEPING THE
PRAYING MANTIS
Orin A. McMonigle

BREEDING THE WORLD'S
LARGEST LIVING ARACHNID
AMBLYPYGID BIOLOGY, NATURAL HISTORY, AND CAPTIVE HUSBANDRY
ORIN MCMONIGLE

FORGOTTEN
ORDER OF THE
VINEGAROONS
Orin McMonigle
WHIPSCORPION

www.ingramcontent.com/pod-product-compliance
Ingram Content Group UK Ltd.
Pitfield, Milton Keynes, MK11 3LW, UK
UKHW060225061026
15625UKWH00008B/493

9 781616 462314